RASPUTIN AND THE END OF EMPIRE

by
Martin Kilcoyne

First print of previously unpublished manuscript 9 September 2019.

Rasputin and the End of the Russian Empire, by Martin Kilcoyne.
original: Kilcoyne, Martin. "The Political Influence of Rasputin"
    Unpublished Ph.D. dissertation, University of Washington,
    April, 1961. (300 leaves/600 pages).

Cover background photograph by Isai Ramos.
Inside title page opposite, photograph is in public domain:
    Rasputin seated with Hermogen and Iliodor.

Design, layout, transcription, and publication by the editor.
Created through Adobe Indesign, printed by Blurb.
Other works published by the editor:
    The Memoirs of Martin Joseph Kilcoyne, November 2015.
    The Role of Military Regimes in Iraq. B.R. McBride, July 2016.
    Ad Absurdum, S. McBride (Forthcoming)
    The O'Byrne Mill [Revisited], C. Tuttle (Forthcoming)

There is no legal basis for the reproduction of this work, save by the author's family. This authorized reproduction is taken from a type-written manuscript in the sole possession of the Kilcoyne extended family, and its copyright has been registered with the Library of Congress.
© TX0008859437 / 2019-10-18

dedicated to Patricia

Editor's Foreword

    The visage of Rasputin still strikes many of us with an uncanny sense of the unknown. Admirers may have seen in his feral appearance some distillation of the Siberian wilderness, some essence of the devoutly religious peasants who weathered harrowing winters for generations. To Rasputin's enemies, those deep-set eyes and the disheveled beard instead were trappings of an illiterate man at best, or at worst the outward marks of a man who allowed some evil to possess him. What is unknown here fuels the fantasy, but most of us are so daunted by the expanse of Russian history that we accept the myths as shorthand. What interest most of us have is in the fiction, which as Umberto Eco would say, is "parasitic" on the reality. When I found myself explaining this project to those who saw me preparing the materials for publication, eyes would light up. They would repeat back the motifs of the myth and savor them, until I said something about the primary value of this dissertation, which is that it challenged the mythical elements—most of all that he controlled the tsar.

    What the average person knows about Russian history typically boils down to this one climactic period. Tsar Nicholas II and his family were fortunate to live in a time that afforded them more photographs than any other royal family. Pictures of each character, place, and event mentioned in this work can easily be accessed online, in greater variety and abundance than could be included here. Published accounts of this period are plentiful. This in combination with the regime's attempt to appeal to peasants may also make them seem superficially relatable. While Martin at times narrates their dramas in Shakespearean terms, he is at other times blunt about their shortcomings. Without this context, the amateur historian can easily slip into indulging their impressions. The most common myth in this way follows straight from visual cues: an idyllic royal family finds itself hypnotized by a strange, black-robed monk. Then one learns of the backdrop, how the appearance of this monk at court indirectly (perhaps by larger forces of darkness working in concert) ushered in the infamous 1917 Bolshevik revolution and the subsequent rise of Stalin. The irony is that Rasputin saw revolution on the horizon and desperately did what little he could to help end the food shortages in his last months. In much the same way that this overstory has been cast as the cautionary tale about the ideas of socialism, the story of Rasputin has been superstitiously cast as a Machiavellian tale of a lusty monk, two antagonists in one body. Reading these two traits into his photographs, we see a man made haggard not by his own travels, troubles, or idiosyncrasies, but by a mounting aggressive paranoia emerging from his own contradictions. The further we remove him from reality, the more these contradictions play into a cautionary tale. The Rasputin we see in these pages certainly played the fox, but the more important story is in a delicate game that allowed him to befriend an underqualified field master for a time, before the hounds eventually circled in around him.

For reasons explained in the first few chapters, the young royal couple faced the circumstances of their rule with some desperation. Rasputin did not work his way to the palace through sheer, inexplicable craft or manipulation. He was fifth in a line of what the monk Iliodor later described as "saint-idiots," or spiritual advisors, plucked like wild orchids from their own various obscure origins in the east. "Most of Nicholas' saint-idiots were only too wise," Iliodor explained, "at least in their own affairs, and played the part of fools only in the presence of others. I believe that Matronushka [the first of these] was a genuine idiot."[1] These were seers, court prophets. One was a peasant-wanderer named Vasili Tkachenko, who made mistakes Rasputin would unfortunately repeat without knowing—particularly the use of royal telegrams as credentials to intimidate or brag to others. The rise of the press in the 20th century amplified matters in a way that made Rasputin's role in this position much more public and subject to speculation than his predecessors. This backdrop to the position he held is a crucial part of the context which the author chose to illustrate in terms specific to Gregory himself.

---

1. Sergei Trufanoff (Iliodor), *The Mad Monk of Russia* (New York: Century Co., 1918), pp. 170-178. Martin discusses this in the appendix, p. 287 of this volume.

Editor's Foreword

My primary qualification for publishing this work is not just that I am the grandson of the author, but that I was the only individual with enough morbid curiosity to bring it to light. Frankly, it was a dream to read this mysterious work that was a part of the oral tradition of my mother's family. His passing in 1989, just six weeks before my birth, left me with a life-long curiosity about what was hidden in his work: this dissertation went unpublished, which may have been a contributing factor in the end of his career in 1972, only eleven years after the dissertation was submitted. By strange coincidence Rasputin was assassinated eleven years after the initiation of his relationship with the emperor's family. In a letter to my widowed grandmother, Donald Treadgold lamented that he "was very sad indeed not to be able to be of more help to Martin in the troubles that overtook him. It was a tragedy that his fine dissertation could never get published and that therefore insurmountable obstacles arose in his career in general." Martin was his first Ph.D. student in a long teaching career at the University of Washington. Martin's denial of tenure had a strange timing and no explanation, leading him to pursue a lawsuit for many years in vain. This had the unfortunate effect of politicizing his resume and barring the chance of picking up where he left off.

Our only guess as to why this dissertation was not published is that one Friday in the mid-1960's Martin left a box or folder by the desk in his university office in Milwaukee, which over the weekend was picked up by a janitor and sent to the furnace with the rest of the wastepaper trash. Desperate panic at the time eventually gave way to an unspeakable grief. It is not clear what precisely was lost, or how it prevented publication, but the material available here is quite powerful as it stands. The two drafts he had stowed away in a briefcase (one 477 pages, the other subsequent, and final draft pared down to 382 pages) appear amended with intent for publication, and now we can see its fruition.

You will find some portions that are historical fiction and others where the mention of a source is considered sufficient warning to quote without citation. Some quotes seem plucked out of nowhere, but the most important elements are clearly anchored in broad reading; in some cases a long narrated section is preceded by a large footnote of consulted sources. Only one glaring error showed up in my cross-reading: Martin somehow led himself to believe the death of Rasputin's brother (a critical turning point in the young man's life) was a dramatic scene of drowning, when in reality we know his brother Dmitri was pulled out from *near*-drowning and instead died several days later from pneumonia. For flow and clarity, I altered the passage where he made that claim. I found this in Joseph Fuhrmann's *Rasputin: A Life,* an author I was first led to read because he is the only Rasputin scholar I found who cited Kilcoyne—"interesting and overlooked" is wedged into his entry for the dissertation, in the bibliography for *Rasputin: The Untold Story.* A second error, directly noted by Fuhrmann, is Martin's attribution of Rasputin's teachings to "folk sects and Satanic cults of the Middle Ages."[2] As a devout Catholic who discouraged his children from listening to rock music, Martin may have had sympathy for that view, circa 1961. However, I found no such claim anywhere in the text. This and the mention of Colin Wilson's *Rasputin and the Fall of the Romanovs* (published in 1964, the timing of which must have vexed Martin) underlines the fact that this volume is as different from the dissertation as its title.

From what I have been able to survey of other works on Rasputin, none seem to approach with the same literary quality as this one does. Many seem to be written and published with the mythos in mind from beginning to end, guaranteeing book sales but also turning the subject into a mysterious object. One way this is done is by splitting the story into as many chapters as possible, with dramatic titles that pull quotes or events out of context instead of presenting a holistic narrative. Fuhrmann was able to sprinkle extra details about Grigori's childhood, and yet there is no mention of his time spent as a carriage driver, nor of Encausse and the superstitions he brought to court before Rasputin entered the picture. What is more concerning is the fact this dissertation's unpublished status made it vulnerable to theft. At the bottom of page 54, Martin makes a summary remark that the charge of being a *khlyst* "dogged him" for the rest of his life. Fuhrmann makes a similarly worded statement in the first chapter of *Rasputin: A Life,* leading to an end-note in the back of the book that reads like a copied and re-worded version, paragraph by paragraph, of an entire section which Martin moved from the first chapter to the afterword. In several smaller points Fuhrmann does this again, but it is likely other authors may have done so as well.

As you find at the start of the first chapter, there were many long passages where Martin seems to have dipped into a reverie patched together by his broad reading of intersecting accounts, creatively filling the gaps with indulgent narrations that invite us into the life of Rasputin as if he were the character of a novel. Martin's ability to inhabit this world in his imagination likely contributed to the positive student reviews he had as a professor. Most of this sort of material was carved out from the

---

2. Fuhrmann, Joseph T. *Rasputin: A Life.* (New York: Praeger, 1990) p. 231.

477-page draft, though I have not yet had the means to travel and cross-examine what was taken out or added relative to the archived original. For this work, the introduction and first two chapters had a fair amount of colorful writing removed for the final draft, but most of it is reclaimed here. Afterward, the pacing picks up and the soap opera of palace life is condensed to the more necessary details. However, large portions of what was scrapped from these remaining chapters are reprinted respectively in the Appendices. The primary narrative in this work concerns the irrelevance of Rasputin's life to the inner workings of government; to the extent that palace intrigue becomes so complex and pedantic that one may lose interest, the details provided are intended to show the true mechanisms that were later written off as Rasputin's influence. In other words, he knew about as much about politics as we do.

What I was able to do is review what sections had been cut out and what *felt* (from my vantage point as a lay reader) like it needed to stay. Some of it may be self-evidently speculative, and yet to me it seemed worthy simply for the fact that it grounds the characters with more sense of reality to time and place than the mythology endlessly portrayed in popular culture.

On a side note, and for clarity, the name Anastasia appears several times in chapters three and six, but it should not be confused with the youngest daughter of the emperor who, according to legend, later escaped assassination. In fact, the famed Anastasia is not mentioned at all. These are references to the princess Anastasia ("Stana" among close relations) of Montenegro who, along with her sister Militza, became a Grand Duchess through marriage to a cousin of the tsar. Her first marriage ended in divorce, which is mentioned, but her second marriage was to another cousin of emperor Nicholas II—Nicholas Nikolaevich, a general. Both Anastasias were likely named after the popular Orthodox saint known as a breaker of chains and bearer of healing potions.

Robin Higham and Dennis Showalter credit this dissertation with marking the end of the "Rasputin myth" in academia, because it broke from the received interpretation of sources cited by Bernard Pares. Letters and telegrams, after all, are a form of theater in their own small way: they rely on a great deal of unstated context in the relationship, and often overcompensate for the lack of personal interaction by showering the recipient with exaggerations. Likewise, publicized accounts of the semi-famed survivors of a political disaster are a kind of theater intended to recast one's involvement in affairs. We can see this in the memoirs of our own politicians and public figures. What the contributors to this Rasputin myth seem to do in various ways is read such sources as part of one grand spectacle with a simple moral that can be guessed before knowing the details. When you do this the contradictions must be ignored because they split the narrative. Ultimately, I think what we find is that a *mundane* Rasputin is at least as interesting as the myth. He was a man of the peasant class which was often carelessly stepped on by the regime, and yet Rasputin was unable to convince the tsar that feeding the hungry could save the empire. He was incredibly lucky to reach the position he filled at court, and unlucky as to where it led him eleven years later. The actions Rasputin took, in this light, were closer to desperate self-preservation than a desire for power. His proximity to the tsar is what made him questionable, in the eyes of family, ministers, and eventually the public. But this proximity is precisely what allowed him to survive as a self-styled holy man, without committing himself to a hierarchy or having to hide his sins in the way he had to back home. His greatest sin might have been his failure as a father, and even then he might have defended himself with Jesus' insistence that followers cast all things aside to follow Him. Rasputin was trying to escape the life he was born into; the fact such a life was lived is interesting enough in its own right.

Martin chose to emphasize something others failed to appreciate: this is a story about religion as much as politics. The faith of the royal couple and their contemporaries played a role as much as Rasputin himself. A peasant "Man of God" would (and certainly, it appears, *did*) go through some extended inward experience of generalizing his own struggles with sin and doctrine, and this is what characterizes his personality. Rasputin was no bureaucrat, and neither was he a mentally deranged witch-doctor. His path involved long years of pondering existential questions, in some fashion, just as Martin narrates. From this perspective, it is hardly surprising his secular onlookers found alternative narratives to explain the man's behavior outside of this context. It could be seen as an indication of their own vanity: Rasputin's story was larger than theirs, and they barely entered into his. It is also correct from a religious studies perspective: the myth of the *khlysty* reflects common motifs of heresy, even those associated with witchcraft in the New World. The endlessly repeated myth of a sect that made a mid-summer ritual of eating the breasts of a virgin and having orgies in the forest was created by inquisitors, extracting confessions under torture. Like Rasputin himself, the *khlysty* cult took on the shape of things which traditional adherents of the Orthodoxy lumped together as blasphemy. This label was useful in making deviants look like part of an organized enemy of established order, but it seems that in the tangible sense of recorded history there was no such group.³

In summary, the story Martin tells of Rasputin scales us back from a moralistic framing that catalyzed political upheaval in the first place. Our protagonist, our 'anti-hero' here is a virtual embodiment of the so-called Boethian account of evil, as opposed to the black-and-white of Manichean theology: evil does not prosper, its successes are fleeting, and its origin is in errors. Evil, that is, may not have an essence in itself, but might be understood in terms of how it deviates from what is divine—whether it be forgetting what it's like to be innocent, "falling" from grace as the phrase goes, or "missing the mark" as the common Hebrew term for "sin" (*chata'*) translates. Martin explores this at some depth, the balance of pride with a humbled acceptance of sin. It was a small victory against pride that Martin was able to move on with his life without publishing this work, but quite a shame considering the thousands of books on popular history that have been published with far less care for facts, sourcing, or detail. Overall, I found the work a fascinating dive into the history of Russia, and I hope you will too. I also found no lack of a proper villain in the story: it can be read as an indictment of the evils that occur in secular affairs, seeming at times insurmountable, or inevitable, because of the fragmentation of truth. ▪

---

3. Clay, J. Eugene. "Literary Images of the Russian 'Flagellants,' 1861-1905." *Russian History*, Vol. 24, No. 4. (WINTER 1997), pp. 425-439.

CONTENTS

Introduction:
    In Search of the Historical Rasputin 1

I   The Siberian Homeland 7

II   A Cry of His Own 23

III   The Road to Court 53

IV   Nicholas II: "What am I to do?" 73

V   Alexandra: Lady MacBeth 83

VI   A *Starets* Ascending 95

VII   Politics, War, and Revolution 151

VIII   The Making and Unmaking of Ministers 181

IX   Where the Road Ended 245

    Appendices 257

    Bibliography 299

Glossary of Terms

archimandrite - orthodox abbot

brodiaga - tramp

'Black Hundreds' - a self-organized political militia of thinkers and propagandists who advocated nationalist orthodoxy and preservation of autocracy. "Black" likely symbolizing humble priests' robes, or the 'dirty work' of the common man

bogoigskatel - god-seeker

dowager - property/title owning widow (as in "dowry")

duma - Russian parliament, literally "house"

khanate - principality of land once conquered by Mongols

khlyst - "whip," used in reference to "flaggelation"

khlysty - semi-apocryphal sect, "Flagellants"; possibly originating as a pun for two related trends: those who emphasized a mystic relation to the Holy Spirit, calling themselves Christ-believers ("Khristy"), and/or the millennialists who anticipated the coming of Christ's thousand-year reign—*khiliasty* would be a term directly Russified from the Greek term "thousand"

metropolitan - orthodox bishop

muzhik - (peasant) man

okhrana - security

sobor - Slavic term alternately referring to cathedrals or councils

skazateli - bards, or storytellers

starets - "elder," spiritual advisor (discussed on pages 34-40)

strannik - wanderer

selo - village, i.e. Tsarskoe Selo, "Czar's village"

tarantass - long-frame Russian carriage that could be turned into a sleigh in winter by replacing the wheels with runners

tovarishch - comrade

tsarevich - prince, diminutive of emperor, Russified from Latin *caesar*

tsaritsa - queen, similarly

ukase - an edict or "imposition"

valide - "mother," borrowed from Arabic for Ottoman matriarchs

verst - now obsolete unit of distance equal to 1.067km (0.6629 mile)

zemstvos - local elected units of government similar to city or county councils

From the reports of the tsarist secret police who watched Rasputin in 1915:

> 12 February.
> Rasputin was taken by an unknown woman into the house number 15-17 Trotskaia Street . . . At half past four in the morning he returned home in the company of six drunken men (one carrying a guitar). These people remained with him until six o'clock, singing and dancing.
>
> 15 June.
> Rasputin left his native village, Pokrovskoe. Rasputin spent twenty-four hours in Tiumen, in the monastery of with his friend Fr. Martian... Rasputin was very drunk. According to Fr. Martian, Rasputin alone drank about two quarts of monastery wine.
>
> 11 July.
> The wife of the officer, Patushinskaia, came... to see Rasputin. Shortly afterwards Solovieva and Patushinskaia came from the house, dragging Rasputin between them, all three intertwined, Rasputin holding Patushinskaia by the lower part of her body. They played the phonograph through most of the day, he being very gay, and drinking large amounts of wine and beer.
>
> 14 October.
> Rasputin returned home at one o'clock at night dead drunk. On the staircase he argued with the wife of the porter... He said [about a minister]: "He wanted to bury me, but he'll be buried before me."[1]

But Rasputin said of himself:

> "I am a humble peasant, but God has spoken to me, and He has allowed me to know what he wishes. The spirit of God is in me and He will protect me."[2]

---

1. "Rasputin v osveshchenii 'Okhranki,'" *Krasnyi arkhiv*, V (1924), 272-73, 275-76, 280. This periodical will hereafter be cited as *KA*.
2. Catherine Radziwill, *Rasputin and the Russian Revolution* (New York: John Lane Company, 1918), p.95.

Introduction:

In Search of the Historical Rasputin

Mention the name Rasputin and you get stock responses. Since the first World War many hastily written books, newspaper articles, and motion pictures have told the world about his strange deeds and the grisly tale of his murder. H.G. Wells, who had a knack for scenting what amount of spice his readers wanted, devoted more space to Rasputin than to Lincoln or Joan of Arc in his immensely popular *Outline of History*.

Rasputin's life was the raw material out of which myths were made. Aside from the obvious details about sex and murder there was in this Gothic tale a hint of dark and bloody deeds, about him there floated an intimation of things of the nether world. The spell of his personality, both compelling and frightening, was irresistible, and it fascinated western readers who had been previously attracted to the bizarre figures of Peter the Great and Ivan the Terrible. Finally, the Rasputin affair took place against the background of a tragic theme: the collapse of a great empire amidst war and revolution.

Romantic writers and readers contributed their share to the encrustation of legend, but some publicists had practical reasons for their presentations. When the government of the tsar fell, Allied leaders found themselves in a difficult position. Russia had been a firm and useful ally. Lloyd George and other statesmen decided to hail the revolution as something beneficial for the Allies because it swept away a decrepit regime that had prevented the nation from making its maximum contribution to victory. But in order to make this excuse sound more convincing the Allies had to thoroughly blacken the reputation of the former government. This minimized the effect of the Germans' boast that they had won a military victory on the Eastern Front and knocked one of the Entente Powers out of the war. The Allies tried to say that tsarism fell because it was rotten, not because it was beaten by the Kaiser's Germany. There are obvious contradictions in this explanation, but it was used nevertheless.

By this time many different interest groups had levelled their charges against Rasputin: he was a confidence man, a traitor, a debauchee, a Flagellant, the real ruler of Russia, the lover of the empress, the father of the heir to the throne. The list of accusations became almost endless. The well known and villainous figure of Rasputin was then made synonymous with tsarism. Some of the excesses of these viewpoints were toned down later, but many of the unwarranted beliefs that went along with them remained.[1]

The short-lived democratic Provisional Government of 1917, before it was crushed by the Communists, began the process of presenting Rasputin to the world. This government apparently hoped to attract support among moderates and even conservatives. It was impossible to appeal to the second group with any kind of political argument coming from the Liberal or Socialist Left. But it could have been attracted or at least neutralized by other kinds of arguments. Rasputin was depicted in the worst possible light and then his name was linked with tsarism. He was shown to be a thorough charlatan and a moral leper—and a typical creature of the former government. This stress on the moral issue touched on values that both conservatives and moderates took seriously. Therefore, tsarism and Rasputinism were to be damned together. When the Communists seized power they continued this policy, adding a few of their own embellishments.

In this way Rasputin entered folklore and history. To this day commentators have had a hard time knowing what to say about the real meaning of his life. Taking note of Rasputin's rise from humble origins to a suspected position of power, one historian proposed that he represented in the government the start of a ministry of all talents. In the 1920's an

---

1. A typical statement of the stylized facts about Rasputin's power is found in P.N. Miliukov *et al.*, *Histoire de Russie* (Paris, 1932-33), III, 1250. "The tsarina followed his counsels blindly, not only in her family life...but also in the domain of general politics. A camarilla surrounded him, and his help was the way to office."

investigator in the U.S.S.R. stressed that Rasputin was a traitor, but from the point of view of the Soviet government the charge was irrelevant. The present edition of the Soviet encyclopedia in a brief entry refers to him as a horse thief. A prominent Marxist literary critic of the 1920's, D.P. Mirsky, dismissed him as "merely the last phase of a long process of degeneration of the old ruling class."

A glance at the literature on Rasputin indicates some of the reasons for the difficulty in understanding him. For one thing there have been few serious attempts to gather and present facts about him. His contemporaries rushed into print with books quickly thrown together, as if they thought that the incredible figure of Rasputin would soon vanish from public interest. Among the secondary works, René Fülöp-Miller's *The Holy Devil* (full citations on this and other works can be found in the Bibliography) is quite good, considering it was written before many useful sources were published. He was one of the few writers who fully understood that if intelligent observations were to be made about Rasputin's activity in 1915-16, an attempt had to be made to reconstruct his early life, admittedly a difficult task given the scanty information available. But Fülöp-Miller's investigation was rudimentary and haphazard in places. His use of some poetic license probably went beyond what most historians would say was sufficient for illustrating a point or making intelligent guesses when trustworthy information was almost absent. Alexander Spiridovich, in his two books, made an honest attempt to put down the facts as he knew them, but his old-fashioned chronological method prevented him from seeing trends in events or from making some generalizations. Perhaps the most disappointing works are those by Rasputin's daughter. They appear to have been hastily done by a ghostwriter who worked from notes made during personal interviews with her. Among the sources there is at least one fraud—the so-called journal of Anna Vyrubova, written, however, by a clever author who had an intimate knowledge of court life. Since all these works were written the presses have continued to turn out bad books about Rasputin. However, the British novelist and critic Colin Wilson a few years ago [1964] wrote a surprisingly good work, despite its errors. The correctness of his views and his insights come not from a sophisticated historical methodology or a wide range of readings but from a mildly sympathetic attitude and a determination to avoid banal suppositions.

The student of the Rasputin affair has been hard pressed to know where to begin looking for reliable material. There has been a tendency, started by the historian Bernard Pares, to cling to one large primary source: the letters of the empress. Taken by themselves these give the impression that she was indeed the co-ruler of Russia. Near the end of his memoirs Pares said that by comparing the dates of certain events with the dates of the letters of the empress in which she passed on Rasputin's words, we can see that Rasputin ruled Russia. After seven years of reflection he wrote *The Fall of the Russian Monarchy* and in it he watered down this simplistic *post hoc ergo propter hoc* notion. It was easy to accept the apparent meaning of these letters at face value. It was easy to be convinced by the demands, exhortations, and trumpeted advice—and most of all the frequent references to Rasputin—that the empress and her friend were all-powerful. However, much careful reading of the replies of the emperor, along with other sources, raises serious questions about such an interpretation.

In addition to the scanty and unsatisfying literature on the subject, there were few reliable contemporary observers. This was not entirely the fault of the men who participated in events; the isolation and secretiveness of the court contributed much to this lack of first-rate witnesses. The imperial tutor Gilliard probably made the fewest errors in his modest attempt to describe what he saw at close range, but even he made extraordinary misstatements at times. Paul Miliukov, a successful and competent historian, on the subject of Rasputin often failed to rise above the level of unintelligent court gossip written down in dozens of works by minds much inferior to his.

It is one of the purposes of the present book to consider the influence that Rasputin wielded in the government. It has been thought that this influence was considerable and that it was exercised in the hiring and firing of ministers. Politicians as far apart in their views as Rodzianko and Miliukov claimed that at one time Rasputin practically controlled the state. Many of the misunderstandings of what was happening in 1915-16 are based on ignorance of who he was, where he had come from, and what were his beliefs. To understand his legend it is necessary to understand his outlook and how he arrived at it. For this reason I have attempted to reconstruct a picture of his ideas and the probable way in which he arrived at them.

It is not easy to grasp what he was trying to think, and many things hide his thought from us. Charles Péguy, when facing this kind of biographical problem, noted that nothing was more mysterious in the lives of religious figures than the periods of preparation where they stood on the threshold of life. Although Rasputin is a twentieth century person, he began in the world of the Russian peasant, a world which belonged to the seventeenth or eighteenth century and is therefore alien to us. Finally, his thoughts and ideas were eclipsed by his powerful personality and his vigorous actions.

How should we take him? Was he a messenger of Satan or the mere personification of a decadent class? Sydney Hook once noted with amusement that some people tried to explain away Adolph Hitler by saying he was the scourge of God; others said he was the result of a crisis in production, but both groups were driven into corners by their logic. We need not take Rasputin at his word and call him a holy man, but he was not a complete fraud, as his enemies said. He seems to have had in his part a period of religious reflection and activity. It was marked by a nagging—although not profound—disquiet in which he wrestled with basic questions that had only conflicting answers. He created some kind of synthesis of ideas. His peculiar potpourri was full of ignorance but was laced with wisdom derived from mother wit. His final position was that of a religion of comfort. He made no stern demands on himself, he denied himself nothing and said that those who wanted to follow him did not have to give up all. His main product was solace. One thing he knew about the world: wherever he went he saw pain and sorrow. Why then did people deliberately have to seek out more pain to add to their natural burdens?

Ernest Hemingway once said, "All things truly wicked start from innocence." It is hard to think of Rasputin as ever being innocent, but he was never as wicked as many thought. If the just man is maligned, we think a crime has been committed, but when a genuine rogue is falsely accused few feel much harm has been done. Even if the moral question is left aside great harm can be done—to historical understanding. To grasp the truth of history it may be as important to know the degree of evil as much as the degree of virtue. A villain deserves honest credit too. Only a fair evaluation (if it is obtainable) can clear our vision and prepare it to see the truth.

I

THE SIBERIAN HOMELAND

On November 1, 1905, the Tsarevich Alexis, heir to the throne of Russia, lay ill in the palace at Peterhof, the imperial residence near St. Petersburg. Outside the child's room, in a hushed chamber the emperor and empress stood waiting with a few courtiers, talking quietly in troubled voices. From time to time the parents glanced through an arched doorway to watch the tsarevich tossing fitfully on his bed and uttering sounds of discomfort. Suddenly, at the entrance to the chamber an attendant appeared. Behind him walked a tall bearded peasant clad in dark, almost clerical garb, and well-polished boots. For a moment he stood motionless on the threshold of the large room filled with furniture and artful objects, tapestries, and pictures. His eyes swept the dim interior, landing on the traditional icon in a far corner, faintly illuminated by a taper burning in an amethyst cup. He bowed reverently toward the holy image, made the sign of the cross on his breast, and then with long strides walked rapidly across the room. Ignoring the courtiers, he nodded respectfully to the imperial couple and went to the bed of Alexis.

There he stopped, his intense eyes staring down at the child. Then he sank to his knees, resting his elbows on the blankets. With a slow and deliberate movement he drew from his belt a small gold crucifix that hung around his neck on a chain. Gradually, he raised the cross before his face, grasping the bottom of it tightly with both hands. His head bowed, and in an instant he was plunged in prayer. For several minutes the rooms were hushed, and all eyes were drawn to the hunched figure silhouetted in the candlelight. The visitor's body was rigid, almost like a statue. Only the thin chain draped from his arm seemed alive—it trembled slightly. Although the deep and resonant sounds of his voice filled the silence of the rooms, his words were indistinct. The listeners were aware only of his emotional intensity and the pleading sounds pouring from him.

After several minutes the tsarevich turned sideways to look with tired eyes at the visitor, then in a moment he fell into a quiet sleep. The peasant waited and presently made the sign of the cross over the bed. Then with a soothing, healing caress he ran his strong hands slowly over the face and head of the child.

Before leaving, the visitor stopped and spoke with the parents, offering them words of encouragement about the heir and about the future of Russia. His speech was a strange but fascinating mixture of the primitive and poetic, and in his voice there was a compelling note of assurance when he said he knew the boy soon would be better. The conversation lasted only a few minutes. Then as quickly as he had entered, the peasant was gone.

At night, after Emperor Nicholas II had finished working with state papers, the made his usual brief entry in his diary. However, he ended by writing: "We have met a Man of God, Gregory, from Tobolsk Province."[1]

This was the beginning of the strange and fateful relationship of Gregory Rasputin and the imperial family. But for Rasputin November 1, 1905 was not the beginning but the end of a long road—one that had brought him from his obscure origins in distant Siberia to the court of the tsar.

§

When Gregory Rasputin became famous in St. Petersburg, people always marveled at his rise to prominence. In the space of a few years he climbed from obscurity to occupy a place beside the throne, where it was said, he dominated the state and church. Observers looked on with complete bewilderment when they considered the dimensions of his career and the possibility that such a thing could happen in Russia. After all, he was born in an unheard of village in a remote part of Siberia, a

---

1. *Dnevnik imperatora Nikolaia II* (Berlin, 1923), p.229.

simple *muzhik* who had never gone to school. But in St. Petersburg he had threaded his way artfully through the maze of court intrigues and survived in the highest circles of state and society where even the most crafty practitioners of intrigue often failed. He consorted at the same time with some of the worst and some of the best persons of society, and he lived a scandalous life that would have brought quick ruin to any other man at court. Without much effort he neatly parried and thrust aside all assaults aimed at him. Statesmen, publicists, high ranking churchmen, and members of the royal family, each armed with skills, supported by experience, and bearing an illustrious name and reputation had hurled themselves at him. Yet one after another they fell before him. How could such things happen? Some whispered that Rasputin had demonic powers that came straight from the Devil. Only such a far-fetched explanation could account for his amazing ability to survive. But there were other, more prosaic reasons that can perhaps explain his triumphs. To understand them we have to look at the world that nurtured him, and try to see what his life may have been like in his homeland.

§

Before the building of the Trans-Siberian railway at the end of the last century, a traveler going from European Russia into Siberia had his choice of two routes. The northern—and the most used—began at a point halfway down the long chain of the Ural Mountains that separate Europe from Asia. Here at the headwaters of the eastward-flowing Tura river travellers could embark on shallow-draft boats, or they could take the road that followed the bank, and move eastward into Siberia for about 250 miles to the spot where the Tura then emptied into the Tobol River. Near the place where the waters meet stands the village of Pokrovskoe, its houses scattered along the left bank of the Tura. In this place Gregory Rasputin was born about 1873.[2]

The southern route left Perm (now called Molotov) in European Russia, crossed the Urals, and headed directly toward Pokrovskoe. At the village the two land roads and water route came together and pushed on eastward to the city of Tobolsk.

---

2. There is some difference of opinion as to the birthdate of Rasputin. Some sources prefer the early 1860's, others the early 1870's. A recent author in the USSR uses the earlier date. "Poslednii vremenshchik poslednego tsara," *Voprosy istorii* (No. 10, 1964), 119.

This was the first part of Siberia settled by the Russians, who began slipping into this mysterious land at the end of the Middle Ages. In the mid-sixteenth century, while Spanish and Portuguese navigators and conquerors were sweeping over the New World, the armies of Tsar Ivan the Terrible destroyed their once-powerful neighbors, the Kazan and Astrakhan Tatars who held vast stretches of land on the European side of the Urals. Having burst through the thin crust of Tatar power the men of Muscovy left Europe behind and plunged at once into unknown Siberia. Later, in 1581, adventurers under the Cossack chief, Yermak, who served the powerful Stroganov family, marched across the Urals and began winning the huge eastern empire for the tsars.[3]

On the far side of the mountains Yermak discovered that the Tura River was a natural passageway into Siberia. As a result, this valley became the first part of the new empire secured by the Russians, who within a few years of their arrival fortified it with blockhouses and forts at key

---

3. The following works were consulted in writing this chapter. *Aziatskaia Rossiia* (St. Petersburg, 1914). V.K. Andreivich, *Istoriia Sibiri* (St. Petersburg, 1887-89). S.V. Bakhrushin, *Ocherki po istorii kolonizatsii Sibiri v XVI i XVII vv.* (Moscow, 1928). A.I. Dmitriev-Mamonov and A.F. Zdziarski (eds.), *Guide to the Great Siberia Railway* (St. Petersburg: Ministry of the Ways of Communication, 1900). *Entsiklopedicheskii slovar* (Brockhaus-Efron), (St. Petersburg, 1890-94). Raymond Fisher, *The Russian Fur Trade* (Berkeley: University of California Press, 1943). P. Golovachev, *Sibir* (Moscow, 1902). Great Britain, Admiralty, Naval Staff, Intelligence Division. *A Handbook of Siberia and Arctic Russia*. Vol. I (London, 1918). Maria Gromyko, *Zapadnaia Sibir v XVII v.* (Novosibirsk, 1965). N.M. Iadrintsev, *Sibirien* (Jena, 1896). R.M. Kabo, *Goroda zapadnoi Sibiri* (Moscow, 1949). A.A. Kaufman, *Krestianskaia obshchina v Sibiri. Po mestnym izledovaniiam 1886-1892 gg.* (St. Petersburg, 1897). A.N. Koulomzine, *Le Transiberién* (Paris, 1904). George Lantzeff, *Siberia in the Seventeenth Century* (Berkeley: University of California Press, 1943). M.G. Levin and L. Potapov, *Narody Sibiri* (Moscow, 1956). Gerhard Müller, *Istoriia Sibiri* (Moscow, 1941). M.I. Pomus, *Zapadnaia Sibir* (Moscow, 1956). P.S. Pallas, *Puteshestvie po raznym provintsiiam rossiiskoi imperii*. Vol. 3 (St. Petersburg, 1773-1788). P. Semenov, *Geografichesko-statistcheskii slovar rossiiskoi imperii*, Vols. I, IV (St. Petersburg, 1863-1885). Yuri Semenov, *Siberia* (Baltimore: Helicon Press, 1963). *Sibirskaia letopisi* (St. Petersburg, 1907). I Tolmachoff, *Siberian Passage*, (New Brunswick: Rutgers University Press, 1949). Donald Treadgold, *The Great Siberian Migration* (Princeton: Princeton University Press, 1957).

Travel Literature written about Siberia at the approximate time of Rasputin's era has been useful. See John Bookwalter, *Siberia and Central Asia* (New York: J.J. Little & Co., 1899). Edgar Boulangier, *Notes de voyage en Sibérie* (Paris, 1891). Thomas Knox, *Overland Through Asia* (Hartford: American Publishing Company, 1870). Henry Landsell, *Through Siberia* (Boston: Houghton, Mifflin and Co., 1882). Jules Legras, *En Sibérie* (Paris, 1899). Victor Meignan, *De Paris à Pekin* (Paris 1876). Fridtjof Nansen, *Through Siberia* (New York: Frederick A. Stokes, 1914). M.P. Price, *Siberia* (London: Methuen & Co., 1912). James Simpson, *Sidelights on Siberia* (London: W. Blackwood & Sons, 1898). Marcus Taft, *Strange Siberia* (New York: Eaton & Mains, 1911).

places, including the site where Pokrovskoe would later be located, for here the Tatars had fought hard to protect the junction of the rivers. On the lower reaches of the Tura, a short distance above the future location of Pokrovskoe, the Russians built the strong point of Tiumen on the ruins of a Tatar fort. During the next two centuries Tiumen became a flourishing market town and leaping point for migrants heading east into the new frontier lands.

As the conquerors pushed on across the top of Asia, they left behind in western Siberia a peaceful region that became a country of trappers, traders and peasant farmers. Towns and monasteries appeared, revealing the progress of Russia's civilizing mission in this part of the world. Wherever they went the Moscovites, like their Spanish counterparts, were quick to rear monuments to their faith and enterprise, even in the wilderness.

At the headwaters of the Tura they built the town of Verkhoture in 1598, and a few years later a monastery was raised nearby. Simeon the Just, a boyar who eventually became a saint, flourished there, and after his death large numbers of pilgrims, many of them from the upper classes, came to pray before his relics. In the first half of the eighteenth century the monastery-town of Verkhoture was the main point of entry to Siberia, and its turnpike enjoyed government-granted monopoly of all freight bound east.

A day's journey downstream from the Tura and Tobol junction, Russians built the city of Tobolsk, which rapidly assumed leadership in Siberian administration and ecclesiastical affairs. At the neighboring village of Abalak they erected a monastery, and because it possessed an icon reported to have caused miracles this monastery became the goal of many pilgrims who regarded it as the most illustrious holy place in Siberia. As Tobolsk grew the church bestowed honors on it. In 1621 it became the seat of a metropolitan, making it the fourth ranking church city in Russia. Rasputin would later send icons from both places to the imperial family, and the empress had one among her possessions at the time of her death.

In the later part of the nineteenth century Tiumen was for a few years the farthest eastern railhead of the empire. At the same time it was the western terminus of the vast Siberian river transport system. The town lived in hope that its railway would become part of the Trans-Siberian line, but in 1885 the trunk line was laid down to the south, and after the mid-1890's the northern region began to lose some of its prosperity, Tobolsk in particular suffering from the dislocation of trade. Tiumen, however, managed to retain most of its affluence. Even after this time the two rivers provided a water route. When they froze during the winter heavy traffic moved along the great post road connecting Perm (Molotov) and Tobolsk. This road, constructed during the reign of Catherine the Great in the eighteenth century, followed the left banks of the rivers, and on it the government maintained an excellent mail post and passenger service.

During half the year when the rivers were free of ice, a fleet of shallow-draft steamboats ranged between Tiumen and distant Tomsk. The shipping line was owned by the Ignatevs, a pioneering family whose operations were centered on the wharfs and building yards of Tiumen. In Rasputin's youth more than seventy ships owned by thirteen companies steamed back and forth on the rivers, and the landing at Pokrovskoe was a regular stopping point where the crews took on firewood and water and the passengers bought food from peasants.

§

Even in the early nineteenth century the inhabitants of the region proudly referred to themselves as old settlers, and government publications later classified the district as one in which these old settlers predominated. Although some of them had to plough land nominally owned by the state, these people had never known serfdom. At the time of the emancipation of 1861 there were only 3,000 so-called serfs in the entire province of Tobolsk, but they were serfs in name only since they paid no money to masters. In this land the master knew that he was living in a sea of peasantry, and if he tried to interpret his rights too close to the letter of the law his estates might be looted and burned. And who would help him? The protecting hand of the government was far away in Moscow. Free land made free men.

After the emancipation of 1861 the area was flooded by newly freed peasants, most of them arriving from Grodno, Vitebsk, Poltava, and other crowded places in European Russia. The newcomers quickly took on the aggressive ways of the older inhabitants, who for generations had been accustomed to looking out for themselves, and who as a result were known for their independence and initiative, personal qualities not often found among the oppressed European serfs. Travelers in this part of the world noticed the difference. A short distance upstream from Pokrovskoe was a notorious shoal. In early September when many of the tributary streams suddenly froze overnight, the river level sometimes fell from six feet to two and a half feet in forty-eight hours, catching captains unawares and grounding boats. Locals reacted swiftly to this opportunity. The sound of straining engines and whistles, the shouts of the crew shifting deck cargo back and forth, and the bright glare of the emergency arc lights against the black spruce forests attracted peasants with horses and wagons. Charging the highest prices the market would bear, they took off passengers who wanted to go to the nearest town. The ship (and its angry captain) were left behind. Drivers knew it still would be there when they came back, and they could gather additional fees for helping to pull it off the bar.

These people of the taiga, the swampy evergreen forest, accepted the challenge of their land. Although their soil was often poor and the climate hostile, they lived better than their European counterparts. They let no chance slip by them. The waters near Pokrovskoe teemed with pike; the villagers fished the waters, dressed their catches, and sold them to passengers on the boats. Wherever the ground was tillable, as it was on the high well-drained right bank of the Tura, they developed a profitable agriculture. But whether the land was good or bad they knew how to take advantage of the busy commercial life and trading activity that went on around them.

The economy of the region throbbed with life of many enterprises. Even in this old Russia the Urals were a treasure house of mineral resources, and for generations the mining of iron, copper, quartz, gold, and semi-precious stones gave work to thousands. For instance, south of Verkhoture was the great sprawling complex of the Demidov mines employing 30,000 men. Loggers worked in the adjacent forests cutting and preparing timber to be hauled by hundreds of wagons to the hearths, one of them the largest wood-burning furnace in the world.

Closer to Pokrovskoe the peasants found other opportunities. Tiumen with its forwarding prison was the distribution point for convicts and exiles deported to Siberia, and inside the city were the grim stockades and compounds of the Chief Bureau of Exile Administration and the forbidding structure of the main prison, a three-storied building of whitewashed brick rising in the center of a fenced courtyard. Nearby was the free hostel in which were lodged children and wives who accompanied some of the exiles. At times the prisons were crowded with inmates, averaging more than 17,000 every year during the revolutionary disturbances of the 1870's and 1880's. Throughout the last twenty years of the century, however, free settlers or migrants in search of new lands far outnumbered the exiles; almost a half million passed through the city on their way to free or cheap land. Finally, there were the religious pilgrims. Every year large numbers of them, individuals or groups, wandered between the shrines and monasteries of Verkhoture and Tobolsk. The presence of all these travellers offered a chance of livelihood to the peasants, who sold food, goods, and services at the Tiumen prison and the stopping points for settlers.

Pokrovskoe was not merely on a road between two important cities; it was on the main highway stretching across the empire. For more than two centuries the most important land and water routes connecting European Russia with Siberia went along the Tura and Tobol, passing through the village. Convict barges, steamboats, caravans of freight wagons, marching groups of exiles, the *tarantass* post, government officials and dignitaries, crowds of pilgrims and much larger crowds of settlers—all these passed through the village in a colorful stream of life moving on the road and river in winter and summer.

§

As they went through Pokrovskoe travellers saw a community extending along the Tura a short distance above the point where it flowed into the Tobol in a long sweeping bend. Both rivers were very wide. The first Russian invaders discovered that if they kept their boats at midstream here they would be safely out of range of arrows shot by Tatar bowmen on either shore. Close to the bank the road followed the sinuous courses of the river as it made a great curve, and the houses of the village were placed here and there for a considerable distance along the side of the road. Many of the inhabitants had individual farms outside the main section of the town, and it was hard to tell where the community ended and where the outlying farms began. But the steamer passenger, who was in sight of the place for almost two hours while the boat pushed against the clashing currents, could locate the center of the village by the presence of the tall gold-colored cupola and high white walls of the church of the Protection of the Holy Virgin that dominated the plain and forest; sometimes at night the presence of the village was revealed by the sound of bells coming from the church. The gold dome, the white walls, and the bells were all proud symbols the inhabitants displayed to show off the prosperity of their town. Visitors were impressed by these things and by the countryside itself. The French astronomer Chappe d'Auteroche in

1761 commented on the rich pastures and cultivated fields around Pokrovskoe.⁴ About the time of Rasputin's birth an English traveller named Charles Wenyon observed that "A fine country lay around us as we steamed up the Toora River. Farmsteads and villages were continually in sight, surrounded by cultivated fields and tracts of meadowland on which large numbers of cattle and horses...were grazing."⁵

Because the junction of the two rivers and the two roads seemed to be the natural site of an entrepôt, the district was settled shortly after the Russian conquest. In those days the church was a powerful colonizing agent that bought lands and settled peasants on them. In 1641 the monks of the Znamensky Monastery of Tobolsk fell into a dispute with the governor of the province concerning jurisdiction over the lands near the village. These valuable tracts were the subject of a long-drawn litigation that progressed upward through the layers of government until the tsar himself had to be called upon to render justice in the case. After securing the lands, the monastery erected the church of the Protection of the Holy Virgin on them (changing the name of the village from Palkino to "Pokrovskoe," a reference to protection).⁶ The village became the center of administration of its region, and its church displayed a popular Savior-type icon given about 1750 by cossacks who said it had "not been painted by human hands," according to the traditional and formal way of describing an icon believed to have miraculous powers.⁷

§

The Siberian post service was an efficient and speedy way of sending mail and passengers across the wide reaches of the Russian Empire. Government officials, who had priorities that gave them the right to preempt facilities, could sometimes cover 300 miles in a day. Stations were located at least twelve miles apart and at each one relays of fresh horses, wagons and carts were kept ready for use at any hour of the day or night throughout the year. Peasant homes along the way served travelers by

---

4. J. Chappe d'Auteroche, *Voyage en Sibérie* (Paris, 1768) I, 88-89.
5. Charles Wenyon, *Across Siberia on the Great Post-Road* (London: C.H. Kelly, 1896), pp. 222-23.
6. P.N. Butsinskii, *Zaselenie Sibiri i byt eia pervykh naselnikov* (Kharkov, 1889), p.134.
7. P. Semenov, *Rossiia. Polnoe geograficheskoe opisanie nashego otechestva. Zapadnaia Sibir* (St. Petersburg, 1907), VI, 394.

providing food and lodging, and many peasants maintained their own posting establishments that competed successfully with the government transport system. The American journalist George Kennan, who was in Siberia in the 1880's, reported that the so-called free post operated by the inhabitants was superior to the government one; it had better horses and its enterprising managers offered more efficient service.[8] The German scientist Adolph Erman, who stopped in Pokrovskoe in 1823, commented on its first-rate lodgings.[9]

Most peasants entered the service on limited engagements when they thought the rewards were good enough to divert them from farming or other pursuits. When opportunity appeared they were free to make their own arrangements with travelers. In this way a group of part time drivers was created among the inhabitants of the villages along the highway.

When he was not working the land Gregory made his living as a driver, one of a wild brotherhood of the road, always on the move, envied and respected by some for their free-wheeling ways, but feared by honest citizens and officials who resented their privileged status and their boisterous celebrations that might erupt at any station whenever a number of them came together. At such moments they drank, danced, fought, got into trouble—and then rode on. Their craft demanded an unusual combination of courage, endurance and good judgment.

The life was hard. In times of storms the post did not run and the driver could find no work, but when the weather cleared he worked almost round the clock to move the backlog of standard travelers. In summer his wagon ran through clouds of dust over pitted roads. At the stations he lounged about, eating slowly, talking and drinking much, enjoying the camaraderie of friends and putting off the moment of departure. But once underway he dashed ahead recklessly to the next stop, hoping to end the ordeal as soon as possible, often standing so that his tired body would not be bounced endlessly up and down on the hard seat. At night he gulped a few quick drinks and dropped wearily onto a bench for a few hours of sleep. Then in the morning he was off on another day of driving. In winter the snow that covered the road was frozen and the runners of the sled raced over the smooth surface. A traveler wrapped in piles of blankets might doze peacefully as the sled skimmed along, but the driver sat in the open with the cold air beating his face.

---

8. George Kennan, *Siberia and the Exile System* (New York: Century Co., 1891), I, 122-24.

9. Adolf Erman, *Travels in Siberia* (Philadelphia: Lea & Blanchard, 1828) I, 311.

A driver had to be shrewd about many things great and small, for life was filled with moments of decision and danger. He had to know when it was safe to drive out on ice newly frozen over streams. Violence might lurk along the road at the crest of a hillock, or at a sharp bend, or wherever the road passed through low trees or brush, the favorite hiding places of criminals who had escaped from prison barges or marching convict gangs. Always desperate and hungry and afraid to go near settlements, they waited to ambush any wagon that slowed down. So even the approach of an insignificant hill might pose a problem for a driver. Should he let the team conserve strength by pulling up the hill slowly or should he rush past the summit to avoid the possible danger?

Moments of decision came suddenly. We can picture Gregory before such a situation, relaxing and drinking with other drivers when an impatient traveler walked up to him and made a generous offer for his services. In a few minutes Gregory had to clear his head and begin to weigh facts carefully so that he could reach a right decision. If he reasoned incorrectly then his life and the lives of his passengers could be lost. Tempted by the bonus, he walked out into the open, sniffing the air, feeling the force of the rising wind against his hand. Then he cast a knowing eye at the grey clouds, studying them thoughtfully. When he had made up his mind to risk the trip he ran into the house and a few minutes later emerged in the yard, leading the team of three horses from the stalls. Jauntily perched on his head was the driver's flat topped hat with its rakish brim and gleaming metal badge showing the imperial eagle of the government post service. A similar badge was strapped on his left arm. He wore thick leather gloves to protect his hands from the hard hempen reins, and around his neck he had draped the harness bells that jingled lightly as he struggled to push the team into the wagon traces.

While he worked quickly to hitch the horses, the passengers wrapped themselves with care in large sheepskins. Then he made a last minute check of the rigging. Satisfied that it was prepared for the hard trip, he jumped onto the seat and shouted to the horses who in an instant leaped forward to a full gallop down the village street. with shouting children and yelping dogs keeping pace for a short distance. Soon the wagon was alone, racing down the forest-lined road under the lowering sky. After another anxious glance at the scudding clouds that were growing darker every minute, he began to turn questions over in his mind. "Will we beat the storm? If we do there'll be a warm stove and some drinks. If we dont..." He knew about the fury that came in such storms when the icy gale roared and drove crystals of snow over the ground, scouring the land, and freezing to death any living thing caught in its path. Into his mind crept the memory of terrible stories of people trapped in the blizzard. Later they were found—driver, wagon, passengers, horses—all covered with snow so that they formed a statue in white stone, standing in place where they were at the instant death struck. His mind jumped back to the present crisis when he heard the howl of a starving wolf coming from the woods. He snapped the reins and to the team shouted loud curses mixed with affectionate pleas. The harness bells rang in a continuous stream of sound, and from the horse's nostrils shot clouds of vapor that formed a cover of ice droplets around their faces. They could smell the coming snow and could sense the growing concern in Gregory's voice as he yelled to them; they responded by pounding down the road, feeling death was running after them.

Later in life, in St. Petersburg, Gregory recalled that in travels in his Siberian homeland he had seen other things that had educated him and filled him with wonder and excitement. He remembered the highways near Abalak and Verkhoture crowded with thousands of people streaming in the direction of the great monasteries, where they gathered for open-air celebration of the liturgy, when scores of priests and well-known bishops clad in full vestments led the processions of faithful who sang and prayed as they walked. He recalled a still summer night when, beside the river across from Verkhoture, he watched the opposite bank high above the Tura where crowds carrying thousands of lighted tapers moved like some large and beautiful serpent slowly around the walls of the shrine. Then there were times when he drove to the Ural mines and foundries, heading toward the dusky orange glow in the sky that came from the fires in the forges and furnaces. His nose was stung by the heavy air, which was filled with acrid smoke and the pleasant smell from wood and charcoal fires. There were nights when the sky burned with another fire—the aurora, pulsing and waving with green and gold phosphorescent lights. There were exciting days in the summer at Tiumen, from 20 June to 20 July, when the annual fair filled the city with crowds of visitors. The wharves along the Tura were packed with steamers unloading cargoes that were taken to the fair or to the railway yards that were now busy with trains moving in and out behind engines whistling and puffing billows of steam. In front of the brick station he watched the uniformed officials moving about with an air of nonchalant importance. Although he was familiar with the telegraph (a line passed through Pokrovskoe) and he was impressed with the messages which went back and forth over the wires, he was especially enchanted and envious when these officials went to their telephones to talk to Moscow and St. Petersburg and other far away places.

Two different beings fought inside Gregory, each trying to capture and dominate his personality, but neither succeeded. First of all, he was the cautious peasant. He was close to reality and was not deceived by mere words or appearances. He worked and sweated to get his bread from the land; he knew how to accept stoically the blows fate dealt him and how to struggle and endure. Secondly, he was the small entrepreneur. He did not believe he had to live and die in the same state into which he was born. Inside him rose the urgings of a restless energy that drove him to seek opportunities to rise in the world. He was alert and ambitious, with an inquisitive intelligence enabling him to make a swift evaluation of the things he heard and saw in his travels. Curious about the world, he was eager to examine new ways of thinking and acting. He was accustomed to gathering facts. weighing them, and making a decision that shaped his destiny. In the rough and tumble of bargaining and haggling he learned about people and how to judge them and their abilities. But he lived with hazards. His path was a narrow one; on one side were the tempting cliffs challenging him to climb to quick success, and on the other yawned the chasm of failure into which he might fall. He could afford few mistakes; he had to rely on a sure instinct and the quickness of his wits to lead him well and to tell him how to exploit the chances that fortune cast his way.

He was cautious but daring. He was capable of placing complete trust in a crony but was always suspicious of the motives of some of the people near him. At one moment he was shrewd to the point of cunning and the next he was unbelievably naive and childlike, but he lived and prospered and made his way in the world despite the obstacles in his way.

His condition of life educated him and gave him an outlook on the world and a system of values. The post road and the activities of the Tobolsk-Tiumen-Verkhoture triangle were his school, and there he learned about the Russian Empire and its people. As a boy he had listened to the pilgrims, officials and migrants who drifted through Pokrovskoe, leaving behind stories and descriptions of the world that burned themselves into his receptive consciousness. Speculation and analysis did not interest him; he wanted only simple concrete images that revealed to him the essential shape of things. As he wandered about and observed the world, a picture of the empire began to form in his mind. The vision was partly the result of his own unique perspective; it was crude and not in perfect focus, many of the lights and shades remained to be filled in but never were. Still, one might find his vision was basically correct, and the proportions of the picture seemed to remain the same for the rest of his life.

§

The voice of the small entrepreneur was strong inside him. But the voice of the peasant proved to be more powerful. He never became one of those rich Siberian merchants who, despite humble beginnings and lack of education, rose to wealth and importance through ability and hard work. From the start he heard a call, soft and distant at first but becoming louder and more insistent with years: "What of the heart? What of the spirit?"

THE SIBERIAN HOMELAND

II

A CRY OF HIS OWN

There are mysteries about the early life of Gregory Rasputin. If we peer into the dark places we find that the facts are obscured in shadows made by him and his foes. Later, when he won a notorious reputation in St. Petersburg, his enemies created a biography showing that from the beginning his nature bore the weight of tainted flesh, a version that became standard in books written about him after his death. Friends, on the other hand, tried to show that once he broke away from the wildness of his youth he became a vehicle of goodness and virtue. To them, he began life as an ordinary man, a humble peasant who was not aware of the extraordinary powers within him, and therefore, not awakened to the signs of his own destiny. Finally, he discovered a purpose in life but only after he had wasted many years pursuing false goals. Although he was continuously defamed, this version said, he went on to win acclaim and honors.[1] To some he was a sinner, to others a saint, but he was neither as evil as his enemies believed nor as good as his supporters hoped.

Many of the incidents in the official biography are presented in a setting that is portentous and apocalyptic, as if he were following an ordained path and nature was cooperating in marking his progress by displays of phenomena—lightning shooting from caves in the earth, ghostly

---

1. For views on Rasputin's life see the following: Boris Almazov, *Rasputin i Rossiia, istoricheskaia spravka* (Prague, 1922). J.W. Bienstock, *Raspoutine, la fin d'un régime* (Paris,

voices born suddenly from a rent sky. Gregory was usually more modest than his friends, but he conceded that mystic signs accompanied his progress. His one attempt to cooperate with publicists came with the tale that in his youth when he was working in the fields one day he saw an apparition of the Virgin. He claimed this was the beginning of his career as a holy man. But he later embellished and altered the story so much that he succeeded only in demonstrating that he had concocted the incident after he had arrived in St. Petersburg.

§

We can see that his biography is similar to those of hundreds of semi-religious figures, both charlatans and honest men, who sprang from obscurity and became well known and controversial. We soon realize that he was presenting us with a crude imitation of a special literary genre, the saint's life. This had once been a popular form of reading and edification, and a source of instruction for the faithful; in Russia its popularity survived into the nineteenth century. Some of the great Russian novelists of that century were strongly influenced by the tales of saints' lives. This was the only kind of biography Gregory knew so he was bound to use it, especially since he believed that a religious experience was the most important fact of his own life. Today when we read the life of a saint we know we are confronting a highly conventionalized form of biography in which the ideal of virtue outweighs an accurate presentation of details. The genre took two forms. First, the official one that cataloged miracles

---

1917). Rene Fülöp-Miller, *Rasputin, the Holy Devil* (Garden City, Garden City Publishing Co., 1928). Gabriel Gobron, *Raspoutine et l'orgie russe* (Paris, 1930). August Lescalier, *Raspoutine* (Paris, n.d.). Heinz Liepman, *Rasputin. A New Judgment* (London: Frederick Muller, 1957). Gilbert Maire, *Raspoutine* (Paris, 1934). Victor Marsden, *Rasputin and Russia* (London: F. Bird, 1920). Lucien Murat, *Raspoutine et l'aube sanglante* (Paris, 1917). Charles Omessa, *Rasputin and the Russian Court* (London: George Newness, 1918). W. Le Queux, *Le ministre du mal* (Paris, 1921). Maria Rasputin, *My Father* (London: Cassell & Co., 1934). M.V. Rodzianko, *The Reign of Rasputin* (London: A.M. Philpot, 1927). George Sava, *Rasputin Speaks* (London: Faber & Faber, 1940). A. Simanovich, *Raspoutine* (Paris, 1918). Alexandre Spiridovitch (Alexander Spiridovich), *Raspoutine* (Paris, 1935). Otto von Taube, *Rasputin* (Munich, 1925). Sergei Trufanoff, *The Life of Rasputin* (New York: The Metropolitan Magazine Co., 1916). Colin Wilson, *Rasputin and the Fall of the Romanovs* (London: Farrar Straus & Co., 1964). T. Vogel-Jorgensen, *Rasputin* (London: T. Fisher Unwin, 1917). Paul Vinogradoff, "Rasputin," *Encyclopedia Britannica*, XXXII (1922), 249. "Rasputin," *Bolshaia sovetskaia entsiklopediia*, XXXVI (1955), 61.

and virtues in a wooden fashion. There were almost no details about childhood, and the ascent to sainthood was passed over. The story really began with the arrival of the state of perfection. Second, the legendary account handed down by oral tradition. This was more dramatic, was filled in with small psychological details, and stressed the path to perfection and the difficulties that attended it. The bards (*skazateli*) who were the keepers of the legend always mentioned something about the cost of the struggle and did not try to dismiss it with a few pious phrases. This feature attracted Gregory to the spoken version, the only one he had easy access to until he learned to read and write. In his own discussions he heavily accented the anguish and pain of the ascent.[2]

There is much useful information in the lives hidden under the artifice and fantasy and the relentless tale of holiness and miracles. Just as the anthropologist has learned to read the outlines of truth in the creation of myths and folk tales, so the historian can use the official hagiography to get the drift of facts of Rasputin's life, despite one important drawback: the old scribes were sincere in their artifice, Rasputin sometimes was not. We know that many details in his story must be connected with the truth, but we are hesitant to stake much on any single detail. Some error, no matter how carefully we sift and judge, is liable to go along with the core of truth. We know about his background, and we have an approximation of the facts in his story, so perhaps we can discern the path that led him to his view of the world.

§

For the people of Pokrovskoe life was filled with endless days of work punctuated by bright moments of idleness or wild revelry. A few pleasures offered relief from the daily round of toil and from the disasters that sometimes struck their lives. Good and bad fortune came mixed together and each peasant in his turn discovered that the pain of tragedy could be shrugged off after it receded into the depths of memory. Life was an adventure; there was little time for brooding.

Gregory was different, but at first he did not know why. He could not see that his sharp intelligence permitted him to grasp and under-

---

2. For the saints' lives see: N.K. Gudzy, *History of Early Russian Literature* (New York: the Macmillan Co., 1949), pp. 25-33, 98-113, 238-43; Sergei Zenkovsky, *Medieval Russia's Epics* (New York: E.P. Dutton Co., 1966), pp. v, 7, 13, 15, 23, 33.

stand things more deeply than his neighbors. It quickened his perception and understanding so that they marveled, and said he had God-given gifts, for when peasants faced the new and unexplainable they turned to religIon for answers. In the beginning he was not sure they were right. But for him there was no balm of forgetfulness in a cheerful acceptance of things as they were; he could not endure in a spirit of resignation the unending cycle of good and evil. Instead of dulling his sensibility, each of life's afflictions increased his capacity to feel pain and his desire to recoil from the aimlessness of existence. Tragedy seemed to speak to him with special meaning, as if it carried a message, or perhaps a warning. Was an evil fate stalking him, waiting for a chance to sting him unsuspecting? On the contrary, every instinct seems to have told him life was not a deceiving viper. Why then did it visit him with troubles? Perhaps, after all, evil and suffering were accidents without any meaning. But only a pagan could believe that. Or perhaps they were punishments for sins. More likely they were God's way of trying to attract attention, to awaken him from a sleep of unknowing.

At night he dreamed he heard a tolling bell; he could not tell where the far-off sound came from nor could he guess its meaning, but it seemed to be summoning him. He was being awakened. But to what? He listened, he was alert, but how did the Lord speak to an illiterate peasant? Eventually he came to suspect that a heavy responsibility rested on his shoulders; he had to find out why he was marked and made different from other men and why he was being called.

Years later he believed he could see in retrospect how the destiny prepared for him was made clear. In conversation with friends he let his mind reach back into the past to recall important incidents that contributed to his growing understanding of who he was and what he was meant for. He said they drifted past him like milestones in a dream which grew more vivid as he approached consciousness.

He thought the first intimations of his destiny came in 1885, when he watched his brother Dmitri wash down the Tura River. They were standing in a shallows pulling in nets holding a few small pike. They were laughing and splashing water on each other. Suddenly a strong current swirled around them. They fought to break its grip and to escape to the safety of the shore. When Gregory crawled onto the bank exhausted he turned and his eyes were filled with the horror of Dmitri's desperate struggle. By the time he was pulled to safety it was too late; the boy died of pneumonia and within a few days all that stood in his place was an empty sorrowful silence. His parents stumbled about the house with glassy eyes, but in time the rhythm of life took over and the deepest part of the pain seemed to go. This could not be said for Gregory. He always remembered the terrible scene on the river; he lived it over and over again in his thoughts. Later he named his own son after Dmitri. But for a long time one thing obsessed him. When death laid its withering hand on the river, both boys were within a few feet of each other, both had fought the same evil force of the waters. Why had he been spared?

After the tragedy neighbors reported that they could see in the distraught Gregory a developing ability to read minds. At times he was sensitive to people's moods and feelings and seemed to know what they were thinking. The notion that such a talent existed was ever present in the villages of Russia. Peasants thought that some people could cure spells, others could "bewitch the blood"—that is, stop bleeding from an open wound. A doctor was almost never seen, but the granny midwife with her magic words and potions (in addition to some practical wisdom) was thought to be just as useful.

Sometimes out of weariness he tried to ignore this call which pursued him, and at other times he pondered its meaning through many a day while he sat hunched on a wagon rolling along a dusty Siberian road. What could he do? It was useless to rage against the world, and there was little he could do to change it. He was accustomed to confronting practical problems of everyday existence; he hardly knew how to wrestle with trials of the spirit. The church was his one place of refuge; for a while he could lose himself listening to the singing before the celebration of the liturgy, but then this too failed him. Almost any phrase of the psalmist set him off: "And my soul is troubled exceedingly: but thou, O Lord, how long?" Hesitations and doubts bracketed his mind as he sought deliverance. Gradually the pressures of doubt and anxiety mounted and the days when he was at peace because of forgetfulness or the distractions of work came less frequently, and he found himself filled with wonder and dread as he thought about his predicament. At times he tried to flee the call, but failed; it pursued him everywhere and gave him no rest. Carousals of drinking and dancing provided temporary oblivion, but it still dogged him. Whatever his fate, he concluded, he was sure that he was not meant to be merely another village wizard, casting out spells or advising about the best time to plant crops. Eventually, he admitted that his call was a spiritual one: he was meant to play a religious role.

But where did he belong? In the church they said contemplation and obedience were ways to God. But he hated contemplation and was not meek enough for obedience. There was no place for him among clerics; an illiterate peasant could not become a priest or monk. But outside of the priesthood one could not live a holy life. There were dangers, and the life of the spirit might grow weak, eroded by strains and pressures of the world.

Again there came a time when he felt he had to share his problem with someone who could advise him. A number of the villagers regarded him with suspicion, others with respect, but no one could talk to him about the meaning of his ordeal. When he carried clergy in his wagon he often tried to engage them in discussion, but most of the learned ones were not interested in spiritual affairs, it seemed to him. In addition, they thought peasants could not experience genuine religious emotions. One day he drove to the Abalak Monastery near Tobolsk. In the church he knelt before the wonder-working icon and gazed up at the face of the Virgin holding the child Jesus. He marvelled at the hair surrounding the face of the Mother of God—it was covered with a mass of bright pearls. He observed the details of the two saints painted in the lower corners of the picture. There he prayed for long hours, and although he felt at peace with himself, he still was not enlightened. Later at the monastery he sought the help of a sympathetic Siberian bishop, Meletyi, an unusual hierarch who because of his liberal political views was watched with suspicion by the government. He had wide support among the peasantry who liked his earthy speech and lack of patronizing attitude. For a long time the bishop listened seriously to Gregory's story, then he told Gregory that it might indeed be true that he had a special mission in life, but it was too soon to be certain. The mere presence of suffering in life was not extraordinary. For many reasons God visited such things on all men. He admitted, however, that Gregory's unusual sensitivity was hard to explain. Meletyi then advised him to return home, to prepare himself by thought and prayer, to go about living in a normal way until some further sign was given to him. The interview closed with a warning to Gregory that he had no choice but to accept his gifts, to strive to develop them for whatever purpose God intended. Since the salvation of his soul was at stake it was not for him to reject his destiny.

With the words of the bishop still in his mind he returned home and made an attempt to live quietly while he waited for an indication of what he was meant to do. He met and later married Praskovia Dubrovina, a peasant woman several years older than himself. She bore him a son, and for a while he felt contentment. Then after six months the child died. The tragedy plunged Gregory into depths of sorrow he had never known before. For days and weeks he wandered in drunken rambles, got into fights, and in several towns earned a reputation as a trouble maker.

He set out once more seeking advice, this time going to Verkhoture Monastery where he visited a *starets* (spiritual adviser) known all over Russia, Macarius, who lived in a hut in the forest near the monastery walls. After listening to Gregory, the anchorite said he agreed with the advice given him by the bishop. Gregory's loss had been a sign, a call from God. Once, long ago, the holy Nikon, who later became patriarch, was stricken with a similar disaster and he knew at once it was God's way of disapproving his decision not to become a monk. So too Gregory had received another sign, a warning that he had to begin at once to search more actively and to discover what he was marked for. If in the quest his spiritual gifts developed further, then he might move up to a new status in the church; he would be called a Man of God, a helper and adviser for the faithful who are seeking enlightenment and assistance along the path of salvation. The monk advised him not to try to find his own path by what he called the "inner way," that is, meditation, but by "the outer way," wandering on long pilgrimages in which he would follow the custom of visiting monasteries, theological schools, shrines, and holy places. Gregory had not dared to think that he might aspire to the role of Man of God. But now Macarius, a cleric well-known for his wisdom, had spoken the awesome words to him.

Gregory took these words to heart and for the next ten years he sought the light of truth in that role, going about the country, talking to monks, priests, and seminarians, and listening in monasteries to readings from the Bible and from the lives of the saints. He hoped that somewhere in his travels he would discover what God wanted him to do.

Until about 1903, especially in the years 1893-96, he was often away from home, travelling fitfully about the country for long and short spells as an inner force gripped and released him. Sometimes he was gone for the better part of a year, at other times he hastened away for short visits to the Abalak or Verkhoture monasteries on important feast days. He dropped into Pokrovskoe mostly during harvesting or planting time and after a while would be off again, leaving the family lands to the care of his father and wife. During the last three years of the century he was at home. His wife then bore him two daughters and a son. In his worldly affairs he accomplished little; the ploughland was not always well tended and the wagon service only held its own, but the Rasputins were clever enough to manage living as well as the other people in the village. In these years the rhythm of his life was broken by the haphazard departures and reunions with his family. When he walked into Pokrovskoe after a long absence his friends greeted him joyously and there were noisy celebrations to mark his return. They settled down to listen to his colorful descriptions of the marvels he had seen and heard on the road. He would then enter a quiet time, for a while returning to his old ways, losing himself in the life of the village as he worked in the fields. Then once more the restlessness began to burn inside him, and he felt he had to go on his travels. Finally, the moment of parting would come. His father complained bitterly about Gregory's deserting his work and duties; they argued and hurled angry words at each other. But Gregory's wife stepped between them. Although she did not pretend to understand him, she supported his determination and sorrowfully encouraged him to go if he must.

§

The roving spirit that caused the Russian people to participate in centuries of national expansion across a wide frontier land of steppe and forest in Europe and Asia also set their feet in motion for other causes. From the beginnings of Christianity in their land they took to the pilgrimage. In Medieval times monks read aloud and inspired the faithful with passages from *The Pilgrimage of Prior Daniel*, a vivid account in the common language of a trip to the Holy Land. By the nineteenth century the Institution had become a normal routine of the religious life of many Russians.

Most pilgrims were part of an orderly movement of pious folk who set out every year on spiritual journeys. They looked on the trip as an interlude in life, although some made a vocation of the pilgrimage. Many went on short visits to local shrines, others went on extended journeys taking months or even years as they swung through a circuit of many of the most important national holy places. Their goals were simple: to induce God to grant them a favor, or to offer thanks to Him for favors already bestowed.

Some were driven by yearnings that could not be satisfied with short trips. Even visits to all the most noted holy places in Russia were not enough to end their longings. Some, driven by inner fires, wandered over the Eastern Mediterranean, an occasional one crossing the Caucasus and descending all the way to the monasteries of the Sinai Peninsula in Egypt. Probably with an eye to the benefits of Russian policy, the government provided good facilities at inexpensive rates to more conventional pilgrims bound for the Holy Land. Special excursion trains took them to Black Sea ports where they embarked on ships bound for Mt. Athos in northern Greece, the Holy Mountain whose monasteries were intimately linked in many ways with one thousand years of the history of Eastern Orthodoxy. From there they went to Constantinople and then to the Holy Land where their needs were cared for by the Imperial Orthodox Palestinian Society. Staying at free government hostels, pilgrims could fulfill the dream of every Orthodox who hoped that before he died he might walk in the footsteps of Christ, visiting the places from Jerusalem to Galilee associated with His life. There was time to unfold a special shirt and dip it in the Jordan River so that it could be taken home and saved as a funeral shroud. Then came the crowning moment: participation in the services at the Holy Sepulchre on the morning of Easter.

At some time in his quest Gregory saw that not all pilgrims were alike. Some had become wanderers for strange motives or under strange circumstances. One might be driven by a sudden impulse that forced him to arise from a meal or to abandon a plow in the middle of a furrow in order to begin a life of wandering. Such a man would strip himself of all possessions and set out without hope of ever returning, deliberately trying to be a stranger in other people's lands. He would have no goal in sight, travelling aimlessly, accepting the pilgrimage as an end in itself, a form of holiness that needed no justification. Another might be hard to distinguish from a professional beggar or tramp. If he appealed to the generosity of peasants they were sure to reward him since they believed that almsgiving to the homeless was an act of merit in the eyes of God. Finally, a pilgrim might take to the road to escape into an underground world of wandering men who traveled without passports, evading the police and living permanently in an illegal status. He might have broken a law and taken sanctuary in the world of the religious Wanderers. In some places such false pilgrims abounded. Near the Monastery of the Caves at Kiev, the cradle of Russian Christianity, visited by more than a million persons every year, bandit-pilgrims who robbed visitors infested a nearby park and its dark ravines. They were especially active at the height of the season, the Feast of the Assumption in August.

Pilgrims were mostly crude folk who left behind no memoirs. The monasteries treated them well, but the church was generally indifferent to them. Intellectuals, brought up in the mocking skepticism of Voltaire and the troubled skepticism of Darwin, looked on them as a bother and embarrassment and tried to ignore them. Only an occasional writer like Dostoevsky or Gogol (the latter himself went on a mission to Jerusalem) paid them much attention. For these reasons the world of the pilgrim is not well known. When Gregory entered this world he almost disappeared from view for a long time during an important period of his life. He put on the garb of a serious pilgrim: a long brown shirt drawn in at the waist with a leather thong, a small metal teapot hanging around his neck, and a crude wooden staff in his hand. He journeyed about the country supporting himself by seeking alms and by doing chores at monasteries that rewarded him with lodging and a plate of sour black bread and buckwheat groats in return for his chopping wood and cleaning buildings.

When this part of his life came to an end he looked back and realized that it had been a useful school. He returned from his *Lehr-* and *Wanderjahre* with a feeling that something had indeed happened to him. He had observed and met all kinds of people who were in some way or another interested in the problems that concerned him. Now after listening and talking to them he had a chance to reflect with greater wisdom on such questions. There was some comfort in knowing that there were other men like him, vexed by the same fears. In monasteries he saw good men and bad. He saw that good and evil were intertwined everywhere as if they were part of the same thing or were somehow necessary to each other. And he saw that truth and knowledge were neither simple nor easy to find. But he was learning where he should not look, where the chances of discovering a nugget of wisdom were poor. While he had not found the words that gave him peace and assurance, he had learned much about distinguishing truth from falsehood. If he did not yet know the worthwhile, he could at least better recognize the dross. These were the tares, but where was the wheat?

This voyage in life had brought pain with it. He suffered disillusionment with the institutions that he had once hoped might teach him. For instance, like most Orthodox he thought that monasticism was the center of the best things in the faith and that behind the walls of the monastery one might find the shining models of a holy life. But he saw that vice and virtue both flourished there. He heard the words of Bishop Cyril of Turov who in the twelfth century had written that the monastic life was the only true Christian life. And Gregory knew of many noble men and women, tsars even, who had been tonsured or taken the veil just before death. After his pilgrimages he knew there was no sure sanctuary of holiness. Other men whom he had never heard of have described the emotional crisis brought on by the shock of this discovery and the scepticism it kindled in them. Luther once said he would have paid a thousand gulden not to have seen the decadence of religious life in Rome in 1512.

And Gregory would have agreed with St. Thomas à Kempis that "those who go on pilgrimages rarely become saints."

Later in his life Gregory made a few cryptic remarks about these discoveries, but he did not dwell on them. However, there was a contemporary whom he never met, a monk who wrote under the pen name Archimandrite Spiridon for an advanced reformist church journal in Kiev[3] shortly before the first World War. Spiridon eventually made his peace with the church, but in his youth he found his faith sorely challenged by doubts brought on by what he saw on his pilgrimages. His career was similar to Gregory's in its early stages. Like Gregory, Spiridon set out as an unsophisticated and devout peasant and had many chances to see monasteries at close range. What he saw disgusted him. We know what Gregory's final views were; perhaps we can use Spiridon's analysis to show how Gregory might have arrived at them. He found nothing good to say about Mt. Athos or Jerusalem. He was especially shocked by the rapaciousness of the monks around the Holy Sepulchre, where clerics of various nationalities brawled over money and privileges, and where Turkish soldiers were called in to separate Christians trying to kill one another a few feet from the spot where Christ ascended into Heaven. False relics abounded. Everyone seemed to have pieces of the true cross for sale, monks snuffed out newly sold candles and peddled them over and over again to duped visitors. Extravagant promises were made about miraculous cures reputed to be guaranteed by blessed relics—if the payment were right. Standahl reported similar fraudulence in his Roman Journal of 1829, finding Aaron's rod and self portraits of Jesus hawked in the Eternal City. Spiridon also lamented about what he saw at the monasteries of Mt. Athos in Greece, where he expected to find a tranquil pool of virtue. Instead he heard constant talk of money and was stunned by the dark vices he found among the monks. The entire atmosphere was poisoned by a recent scandal when the Holy Mountain became the scene of a heresy among the inmates of the Russian St. Philemon Monastery. The tsar's soldiers with bayonets on their rifles had charged into the rebellious monastery and dragged out the kicking and cursing monks who were put in chains and shipped back to Russia. In the community the air still crackled with tension and bitterness for many years after this episode. In the first World War the Empress Alexandra, writing to her husband, mentioned the affair with some apprehension, fearing that it might flare up again. In summarizing the effects on him of all these depressing

---

3. Archimandrite Spiridon, *Mes missions en Sibérie* (Paris, 1950), pp. 36-37, 51.

things, Spiridon remarked about the monks: "I was very disenchanted, because, with all their...exploits, they lacked a moral side of life." About Mt. Athos, he concluded dryly, "I never saw any saints there."

§

The resolute pilgrim was so often away from home that many of his ties with the past and with his birthplace were cut; those for whom traveling became a way of life were called a God-seeker (*bogoiskatel*) or more commonly, a Wanderer (*strannik*). If the wanderer developed his spiritual powers and showed an ability to help others he might become the supervisor of the spiritual lives of those who consulted him intimately. Such a man was called an Elder (*starets*; plural, *startsi*; the cult was called *starchestvo*).[4]

When Gregory set out on his travels, the Russian church was already more than one-thousand years old. In many ways it resembled its sister institutions in western Europe, but in other ways it was strikingly different. As a religion of the people it was probably not much more unchanging than the church in Portugal or Sicily; it had kept intact ceremonies and institutions which belong to the days of Bede, Alcuin, and Patrick. Its own historical experience—defending a far frontier of Christendom and its emphasis on preserving ancient practices—had given it a special character which westerners regarded as primitive or quaint. The monasteries, not the schools or the urban churches, were the centers of piety in Russia. The pilgrimage was not a rare hold-over from the remote past; it was a vital part of the religious life of many of the faithful. Around the numerous holy places of the country at certain times of the year huge crowds of visitors congregated; in the woods near monasteries forest monks dwelled in pious and ascetic solitude like the eremites of Egypt or Syria in the sixth and seventh centuries. Frequent religious holidays and ceremonies caused normal business life to be suspended. In one of the large railway stations of St. Petersburg a towering picture of the Virgin made up of colored electric light bulbs glowed down on hurrying crowds of travelers. Religious ceremonies accompanied many state functions from diplomatic events to military reviews. Intense mass piety

---

4. For *starchestvo* see: Igor Smolitsch, *Leben und Lehre der Starzen* (Köln, 1952). Ivan Kologrivoff, *Essai sur la sainteté en Russie* (Bruges, 1953), pp. 377-400. Anonymous, *Otkrovennye rasskazy strannika dukhovnomu svoemy ottsu* (Paris, 1948).

manifested in a public, open fashion intruded on and even pushed aside the orderly pattern of the modern secular world. Although Orthodoxy had lost some of its appeal to the upper classes and in some ways had become a state cult, it was still an important part of the fabric of national life, especially with the people.

If Gregory's future was to be in the church, he realized he would require a special relationship with it because a poor and illiterate peasant could not become a cleric even though he felt he had a vocation to be a holy man. He began to move into the realm of popular piety where there were several institutions to serve him.

The best known was *starchestvo*. If a *starets* won recognition for spiritual powers and showed an ability to help others, he might become the supervisor of the spiritual lives of persons who consulted him intimately about religious problems.

The idea appears in several forms in most of the major faiths of the world which recognize the role of the wise man or guide. In Russia the cult appeared early and many of its practitioners were canonized. It was inevitable that in the time of triumph for royal absolutism after 1700 *starchestvo*, which had certain independent features demanding a privileged and inviolable relationship of Elders and followers, was regarded with suspicion by officialdom. During the course of the centuries hostility replaced suspicion. The Holy Synod—a bureaucratic agency that controlled the church—withheld the honor of sainthood from *startsi*, although it permitted them to exist if they were discreet and unostentatious.

Laymen and clerics who wanted to regenerate the national faith turned to ideas and institutions existing outside the official church in the hope that they could be safe from the interference of the government. But the head of the Holy Synod (called the Over Procurator, the real head of the church) watched jealously at the birth of any new religious ideas or plans. After hundreds of years of struggle, Orthodoxy was firmly under control; new ideas might only disturb the comfortable grip the state had on the church. Despite this difficult situation men appeared who sought to revitalise the guiding and teaching role of Orthodoxy. The modernizing of the Russian state apparatus in the nineteenth century created new secular classes of administrators and merchants who were not as easy for the church to reach as were the classes of old Muscovy. The problem was how to speak to them without arousing the government.

In eighteenth century Europe the flame of evangelical religion had flared up: there was a surge of interest in the religious experience of the individual, a concern with personal piety, and a quest for emotion. One humble pioneer of the trend in Russia was St. Serafim of Sarov, whose canonization in 1903 played an important role in the career of Nicholas II. Seraphim was a practitioner of heroic virtues reminiscent of the eremites of the deserts of the Near East fifteen hundred years ago. At one time he was a *starets* although he did not stress this role. He thought of it as a way to share the spiritual riches of his own life with laymen.

In the early nineteenth century a new church leader arose, Paisius Velichkovsky, who wanted to restore the church as a vehicle of inspiration and guidance for the people. He turned to certain currents of thought from Mt. Athos and to old traditions within Orthodoxy that stressed personal piety and the need of the church to avoid close ties with the state. He hoped that *starchestvo* might bring Orthodoxy and people together. He claimed that every man would be taught to live a life of spirituality. A person who sought this goal had only to subordinate his will to the direction of an Elder who would suggest a regimen of guided prayer, readings, and some meditation. Paisius' solution avoided raising any questions about the relation of church to state, nor did it call for new institutions or rites. The natural elitism of refurbished *starchestvo* protected it from any strongly hostile reaction by the government, but its followers had to be cautious, nevertheless.

Paisius insisted that perfection was for the few, the elect who were able to cultivate deep spiritual feelings that could light the spark of holiness in the hearts of others who would in turn make their piety spread through the church. The followers of the cult gathered at the Optina Monastery near Kaluga in central Russia where they were able to get around the censorship and to publicize their ideas with the help of their protector, Metropolitan Philaret. This monastery became a center of the movement, and in a short time its fame spread all over the empire. From there the cult diffused among the upper classes and in many places, including court and society, eminent *startsi* appeared. *Starchestvo* in its new guise became popular. For almost a hundred years an unbroken line of famous Elders resided at Optina providing consolation and guidance for many distinguished visitors. In the summer of 1878 Dostoevsky spent several days there where he talked to *starets* Amvrosy, who became the prototype of Father Zosima in *The Brothers Karamazov*.

Starchestvo was never formalized. A man entered it not by choice; the call came to him and he was forced to accept it. Sometimes the honor came despite his own desire. The path he followed in becoming a *starets* was poorly defined. Usually a series of strange incidents in his life caused an awakening. After he recognized these as portentous signs or theophanies he began a long period of apprenticeship under an Elder. If he were a monk he might, in some cases, walk alone on the path to perfection. If he finally exhibited great spiritual powers, then he could win recognition as a *starets*, either by his own director who was able to preempt him, or by the people who greeted him with a spontaneous acclamation. For the most part *starchestvo* was a religion of healthy mindedness marked by an air of reasonableness and an absence of mysticism. There was no morbidity or starvation, no chains and whips. The *starets* might have once been an ascetic, or there might still be a corner of his life where he practiced such virtues, but he did not turn this face to the view of most persons consulting him. To be most effective with the largest number of laymen he preferred to deal simply and directly with their problems. Sometimes he would claim the ability to see into the future of the person he was guiding, but this was unusual. He had only to convince people of his superior ability to make wise judgments and to give useful advice about the way to perfection. The *starets* was a minister to those who struggled in the work-a-day world; he sought to help men of affairs in government or business. Austerity and heroic virtues therefore had no use in this method of spiritual perfection.

Ignorance about this cult caused many people to misjudge the role of Gregory at court. Sometimes the relationship of adviser and listener could temporarily assume an extreme form. The bond between them was to be of the most intimate kind; the follower was supposed to renounce his will and to follow the *starets* with absolute obedience along the Way of the Cross, as it was called. However, this was unusual. The follower was usually trained to stand on his own feet, not to learn meek obedience, a technique that might be airtight for a monk who lived away from the world. But Fyodor Dostoevsky, for dramatic reasons peculiar to his own interests and tastes, in his *The Brothers Karamazov* described the extreme form of the *starchestvo*, and many observers whose only contact with the cult was through the pages of this novel thought all relations of *starets* and follower were like this. As Dostoevsky described it,

> An elder was one who took your soul, your will, into his soul and his will. When you choose an elder, you renounce your own will and yield it to him in complete submission, complete self-abnegation. This novitiate, this terrible school of abnegation, is undertaken voluntarily in hope of self-conquest, of

self-mastery, in order, after a life of obedience, to attain perfect freedom, that is, from self.[5]

He added, "The elders are endowed in certain cases with unbounded and inexplicable authority" over the individual. This description refers more to the relation of two clerics, one of whom was trying to reach almost the condition of sainthood.

While *starchestvo* was growing among the clergy, a popular version of the movement flourished among the common people. An ordinary peasant might become a spiritual guide of other peasants; no one ever thought of him as a potential adviser or confidant of members of the upper classes. Such a holy man was generally called a Man of God, a term indicating the less clearly defined nature of the education and status of the peasant *starets*. The goals of the peasant were not always as clearly marked as those of the monks. Moreover, his followers demanded that he be able to help them in practical ways: he had to predict the future, to undo spells and have a ready collection of potent words and formula for relieving pain or curing sickness in man and beast. A peasant Man of God might achieve recognition in a small community or he might be sought out by people in an entire province. But the village clergy, who were favorable to the cloistered *starets*, did not like the Man of God. He usually was in direct competition with the priest, and he often had learned corrupt religious ideas in his travels.

The Man of God fashioned his own rough methods for discovering his calling. He might never progress beyond the stage of a wanderer, and as a result he might spend his life drifting about the country begging food in the villages and eluding authorities.

§

Men of God were sometimes confused with practitioners of another kind of popular piety, the 'Fools of God.' The church recognized the distinction between clerics who were *startsi* and those who were Fools of God, and many from both groups had been canonized. Peasants tried sometimes to imitate both styles but often made themselves into mere degraded or confused parodies of the classical clerical ideal. Furthermore,

---

5. Fyodor Dostoevsky, *The Brothers Karamazov* (New York: Modern Library, n.d.) p. 24.

among the people the Fool of God was often thought of as a superior kind of *starets*, for to achieve this status was to pass beyond *starchestvo* into a higher holiness that brought a man closer to the divine. In official circles, however, the Fool was regarded with even more mistrust than was the Elder.

The idea of Fools came from the New Testament. In the eleventh century several of the great churchmen in the Monastery of the Caves at Kiev became Fools. St. Theodore practiced "self-emptying," and although he was born a noble he wore ragged clothing, worked the fields with slaves, and gladly accepted the contempt of the world. He followed the command of St. Paul to be a fool for Christ by "taking on the form of a servant." The Fool acted and spoke in ways that invited the amused scorn of society. He sought to be insulted and injured, to be mocked and despised by the world. He deliberately lowered himself by renouncing prudence and good sense, by abdicating human dignity and pushing self-abnegation to the heroic edge.

As with the cult of Elders, the Fools declined when the tsars of Moscow consolidated their power over the church. At first they had been accorded certain privileges as rewards for their sufferings which were connected with their humiliation and lowly condition. They could move about without being molested and they were free to lecture and threaten the rich and powerful. They served as social and political critics, aiming their barbs at two hated classes: the brawling nobles and greedy merchants. There was a common belief that when the Fool spoke he uttered words the Lord had given him. West European travelers in sixteenth century Muscovy, Bishop von Herberstein of Austria, and Giles Fletcher from England all commented on the way in which the Fools spoke out against abuses and were neither silenced nor punished. The state might even honor them. The exotic multi-domed church which still stands prominently in Red Square in Moscow was built by an Italian architect on the order of Tsar Ivan the Terrible, who wanted to honor Basil the Blessed, a well-known holy man, a Fool of God who had denounced Ivan and had not been punished for his daring.

Some of Gregory's enemies feared he was playing the Fool in order to take on the protective coloration of the cult, for in such a role he could enjoy immunity from certain kinds of official criticism. They feared he was playing the Fool who, in lecturing and even threatening the tsar, was able to exert unmolested a kind of evil influence over him.

§

If Gregory was going to be a *starets* he had to have a clearly defined faith. People who held a *starets* in high esteem consulted him because he could lead them from their own wavering ground to the firm rock of his belief. Gregory hoped that there was a place where a peasant Man of God might find such a faith.

The forest monks had pointed to the path, but only he could make the journey. As a Godseeker and Wanderer he had failed to find what he was looking for. His wandering was a time when hope was betrayed and expectation led up a road to disenchantment. He realized that one by one the sources of truth had failed him. His choice had been limited because the springs of Orthodox learning were barred to him, an illiterate peasant. He had to rely on the long discussions he had with monks and seminarians when he stopped on his travels. He spent much time passively listening, but he often talked for long hours during which he heard many things that interested him and might be useful to him. He eavesdropped on the discussions of learned men and listened to the readings from some of the works of ancient holy men. He pondered what they said, hoping to find words that could guide and inspire him. But there was nothing.

Once at a fashionable salon, in the presence of a church official (N.D. Zhevakov, Assistant Over Procurator of the Holy Synod) Gregory once explained that he had examined the lives of the saints and holy men but they offered him no guidance.[6] Only a few laymen had won sainthood, and many of those were princes who had been able to withdraw from the world. The stories always began when they had already been successful and put behind them the days of struggle and doubt. He reasoned that there must have been sinners in their ranks because they always said that they had to struggle up to sainthood. This must have indicated that they did not bother to talk about their early struggle because they knew each man had to find his own path, and his discovery had no bearing on the similar problems of other men. There was not one path, there were many. But what was his?

At one time he thought that the truth might be found in holy books. Perhaps the monks, the church, and the deacons were not reading or interpreting them correctly. So he learned to read and write in

---

6. N.D. Zhevakov, *Vospominaniia tovarishcha ober-prokura sviateishago sinod* (Munich, 1923), I, 277-78.

order that he might pour over the holy wisdom to make it yield its fruits. But the didactic writings of the church did not help him much. In the *Supplication* of the wise man Daniel, who wrote just before the long twilight of the Mongol Yoke settled on Russia, the advice was contradictory and seemed to justify opposing ways of conduct. In such counsel there was no firm direction. Nor was there anything to interest him, later on, in the *Life of the Archpriest Avvakum*, the fierce cleric and sectarian of the seventeenth century. Although Avvakum had travelled his own *via dolorosa* from doubt to fulfillment, he could not speak to Gregory, who was repelled by the priest's fanaticism. Avvakum described himself as "a miserable sinner, in churches and houses and at crossways, by towns and hamlets, even in the city of the Tsar and in the country of Siberia."[7] Well and good, but Avvakum seemed to say a man was either a saint or sinner. Gregory could not attain the first state, and he was trying to raise himself up from the second. What his reason sought was a third condition, a place where there was holiness and worldliness together.

He scrutinized the Bible and learned parts of it by heart. When he arrived at a monastery he liked to attract attention by joining a circle of monks and visitors, and there in a quiet setting he would recite long, difficult passages from scriptures. In talking to groups of peasants he repeated word for word stories of heroes and prophets of the Old Testament, and he gave the official interpretation of the parables of the New Testament. These were simple and hackneyed but always pious and correct. Later he was surprised to find himself inspired to add embellishments and his own interpretations. The peasants responded to these bits of fancy; they asked for more. Even some of the monks and the students were highly pleased. But the teachers and leading monks showed concern. They feared originality in peasant pilgrims or holy men, for such things often led to heretical thinking. Gregory learned to save the play of imagination for peasant gatherings. He likely found that the upper clergy responded warmly and were impressed if he performed virtuoso feats of memory and spoke in a reverent way that they thought fitting for a peasant Man of God. In dealing with the Bible, Gregory had taken his first lesson in the need for caution and the necessity of sometimes keeping two faces. Inside him there was still the flickering hope that he might find a unified, coherent faith and a role for himself in the church.

At any rate, the Bible was no longer regarded as a final source of enlightenment. In fact, when Tsar Alexander II came to the throne he or-

---

7. *The Life of the Archpriest Avvakum by Himself* (Hamden, Connecticut: The Shoe String Press, 1963), pp. 43-44.

dered a commission of scholars to prepare the first full translation of the Bible into modern Russian. When they completed their labors the Holy Synod in 1876 published this version in an inexpensive edition. Russians went to the written Bible to check quotations or to follow the parts of it mentioned in church services. It was one of the church's pillars of truth, but it was not a place for the individual to make discoveries. Gregory soon realized he was no longer seeking his own version of conventional Orthodoxy. He was slipping into new modes of thought.

This part of his education began at the Verkhoture Monastery. Those who spent time there became aware of strange happenings that casual visitors did not see, curious things that puzzled Gregory. Some of the monks were jail keepers who watched over other monks who appeared to be prisoners. This was one of several monasteries set aside by the Holy Synod to exile unruly clerics suspected of being tainted with sectarian ideas.[8] The most dangerous heretical monks were sent to the Solovetsky Monastery on the islands of the forbidding White Sea in the Arctic. Near the central Russian city of Vladimir monks were imprisoned in an institution that was spoken of by the clergy in the same way revolutionaries whispered about the dreaded Peter and Paul Fortress in the capital. Attendants at Verkhoture made half-hearted attempts to rehabilitate the prisoners, but they were too busy trying to serve the crowds of pilgrims. As a result, the inmates continued to hold their beliefs and were able to propagandize some visitors.[9] Gregory frequently had reason to go to the monastery on business, and it was there that he may have first heard about the ideas of the sectarian Christians and of their criticisms of Orthodoxy.

The national church no longer had an exclusive hold on the allegiance of most Russians. In the middle of the seventeenth century a schism had rent Orthodoxy and despite persecution the strength of the rebels grew. On the eve of the first World War there were about twenty million of them, but they were split into many groups.[10] Some, such as the Stundists, were rational and moderate in their convictions. Many were exotic and fantastic. Among the best known of the extremists were the Castrators; the Flagellants (*khlysti*), supposedly friends of Gregory; and

---

8. J.S. Curtiss, *Church and State in Russia, The Last Years of The Empire. 1900-1917* (New York: Columbia University Press, 1940), pp. 70-71. Anatole Leroy-Beaulieu, *The Empire of the Tsars and the Russians* (New York: G.P. Putnam's Sons, 1893-96), III, 252.

9. Fülöp-Miller, pp. 18-19.

10. Curtiss, p. 139. This is actually the largest figure given by one authority. Curtiss seems to prefer 17,500,000.

Dukhobors, ("Wrestlers of the Spirit") who were supported by Leo Tolstoi. All these groups experimented with a variety of rites and beliefs, and taken all together their movements formed an interesting treasure house of strange ideas and customs. Eventually Gregory saw these people in the light of his growing awareness that men might go along many different paths to truth.[11] He examined the ideas of the sectarians with an open mind. Gradually, without realizing what was happening to him, he began to drift away from the traditional views of Orthodox learning. There were many other ways open to him.

He never thought he could live without religious conviction. It seemed that all men were believing creatures, and the Russian Empire was a museum of many faiths. He had seen many of them in Tobolsk and Tiumen provinces. To the south of Pokrovskoe, in the Ishim Steppe, lived the Islamic Tatars, nomads who thought and lived differently from the Russians but who clung to their religion with fierce conviction. Not far from Tobolsk he saw tribes of pagans who worshipped idols and the forces of nature, and along the Tura were mysterious villages with secretive people who professed two religions at the same time: publicly they behaved like Christians but privately they were heathens. In the towns of the province were several hundred Lutheran families. When he went on his travels in the western parts of the empire he saw Jews. All these people worshipped with equal fervor, all believed they were right, and persecution had not diminished the intensity of their convictions. Rich or poor, ignorant or educated, they clung to their faiths. But they all disagreed on the nature of truth. What puzzled him were the different cries each group raised when they were asked to speak of what they believed.[12]

§

Many of the saints' tales contained the story of the instant of conversion. But for Gregory there was to be no drama, no moment of inspiration when enlightenment poured down on him to wash away his blindness. Revelation did not strike him with a sudden flash of light. But

---

11. Marie Rasputin, *Rasputin* (London: John Long, 1929), p.43.
12. Many people who claimed to have met Rasputin or observed him tried to write down the content of his ideas, and some attempted to capture his unique ways of expression. A book of his letters—probably edited—was published: *Moi mysli i razmyshleniia* (Petrograd, 1915). For a shrewd commentary on such sources see Wilson, pp. 46-47.

if God had long ago given faith to the heathen prince, Vladimir, who became a saint and founder of Christian Russia, why, Gregory wondered, could He not hear a poor peasant? He waited in the wilderness hoping to experience the moment of awe that came before the dawn of faith—but none appeared. He cried out in his fashion, "Lord, what wilt Thou have me to do?" But there was no answer.

Slowly and painfully he was sloughing off some of the questions that nagged him, but there was still a long way to go. He saw that his faith would not be given to him in a package, ready made for him to take and use the instant he received it. He would have to accept his faith by pieces and patches as he discovered it in his wanderings. The entire fabric could be put together later—he thought. But shaping his beliefs into a consistent whole was a task beyond his powers, and in St. Petersburg, beyond his interests. His outlook was clouded not by fanaticism but by ignorance. It was filled with contradictions and inconsistencies, but it suited his needs. Life had taught him to live with paradox; for a peasant the world always was full of riddles. In the capital he modified his convictions so that he could better exploit new situations confronting him. But at all times he was determined to keep his individual viewpoint and to place on his belief the stamp of his own mind, which at times was muddled but independent.

He understood that the power of the government stood behind the church. He was also well aware that the church was militant and alert when dealing with deviations from its teachings. Therefore, prudence demanded that he not flaunt some of his more unusual ideas in public. In addition, his formal outlook called for suppleness in beliefs and a moderate approach to all articles of faith. There were other reasons for his caution. He felt comfortable as a member of the Orthodox Church, and he intended to remain in it because to leave it would force him to accept pain and sacrifice. He did not think there was any glory in such things. Gregory-as-driver advised Gregory-the-fanatic about the high cost of a martyr's crown. It was not for him; he was a coward, not a masochist. Moreover, he had no liking for any kind of zealotry, his own included—especially when it might hurt him. Finally, his faith was a confused melange of ideas, hardly the thing to suffer for. Only one strong conviction glowed amid the confusion of his outlook: he thought he was guided by an inner voice and that he had a right to hear it and live by it as much as the authorities would let him.

A spirit of liberation filled him when he realized that he might not have to accept the guidance of every written law that bound him to the past. He was free to be the arbiter of his own faith. In some respects Orthodoxy no longer suited his needs; he now felt free to reject the parts of it that did not measure up to standards he chose. It was no longer an infallible guide in all things.

As he thought of his relationship with the church he considered its obvious flaws. First of all, although he revered it and its sacraments and traditions, he disliked it as a bureaucratic agency. This poisoned the wells from which the church drew its life. As a result it was weighted with human weakness and many of its hierarchs were not worthy men. It had become stagnant, lacked true piousness, and too many of its people were interested in place and material reward. The church should first of all be an organization that helped the common people. It should be concerned with the welfare of its members in this world as well as in the next. Gregory noted that the church had once seen its duty in this way. St. Theodore of the Monastery of the Caves said that the monks owed their first duty to the poor. The great Medieval prince, Vladimir Monomakh said in his *Testament*:

> It is neither fasting, nor solitude, nor the monastic life, that will procure you life eternal—it is well-doing. Do not forget the poor, but nourish them. Do not bury your riches in the bosom of the earth, for that is contrary to the precepts of Christianity. Be a father to orphans, judge the cause of widows yourself.

St. John the Almsgiver, Patriarch of Alexandria, had proclaimed: "Those whom you call poor and beggars, these I call my masters and helpers. For they, and they alone, can really help us and bestow upon us the kingdom of heaven."

In the end his religious views were tinged with rationalism, when a streak of mother wit caused him to veer from excesses, but other views represented a retreat into superstition. He thought, for instance, that objects in themselves had miraculous powers.

He did not apologize for his ideas, nor did he debate or hold discussions in order to win converts. To him religion was a private matter. His own was sufficient for him, but he was willing to talk to others provided he did not have to argue with them. One found the truth by oneself; no one could be converted by mere words. He was not arrogant about his beliefs, nor did he feel inferior or apologetic. But he was confident; he found ways to quiet his concern about becoming a religious figure, in spite of his sins and illiteracy. At some time Gregory heard of heretics of old (Gnostics) who said that faith and knowledge were one. The educated thought that as a peasant he could not understand the real meaning of Orthodoxy because he was unable to grasp all of its most subtle dogmas with his intellect. Gregory learned to turn this around: worldly knowledge was not one of the rewards promised by God to those who had faith. It might even be denied them. The ignorant man, in fact, might be closer to God because he held the faith through the urgings of his heart and feelings, and these were better guides than the mind. Knowledge might even get in the way of the seeker. The peasant's heart was close to the spirit, and the spirit quickened understanding. The Apostles were ordinary men, fishermen who made their living with their hands rather than heads. It was Satan, after all, who quoted Scriptures.

Though such ways of thinking helped him to loosen the ties binding him to Orthodoxy, he did not intend to be aggressive in dealing with the problem of authority. He was aware that the power of the government stood behind the church, which was militant and alert in dealing with deviations from its teachings. Therefore, prudence demanded that he not flaunt some of his heterodox ideas in public. In addition, his formal outlook called for suppleness in belief and a moderate approach to all articles of faith. And there were additional reasons for caution. He felt comfortable as a member of the church, and he intended to remain in it because to leave it would have demanded pain and sacrifice. Furthermore, he conceded that the church might be the only ark of truth—for some. If he had an attachment to any religion it was to Orthodoxy. Since he was a Russian he was willing to admit that the national faith had a claim to his allegiance. Like all Russian peasants, he expected Turks to be Muslim and Germans to be Lutheran, and it was fitting for Russians to be Orthodox even if they had reservations about some of the church's teachings.

There were many things he objected to—laws, dogmas, decrees, consistories, hierarchies. These were abominations to him, useless growths choking off the flow of true Christian feeling. But most of all, he objected to the bureaucratic spirit that poisoned the sources of faith and corrupted priests, turning them into place-men interested only in careers.

His faith was full of emotion, of instinct pulling him to the forgiving heart, away from wrath and fury. He looked for the spontaneous and the sincere: the act of mercy done without calculation, generosity extended without nit-picking questions, charity without reservations. These were truly good works, and they were the best part of faith.

Gregory gave whatever he had. When he was a Man of God, before he had acquired some influence, all he had to offer was hope and comfort, which he dispensed as spiritual consolation to the afflicted. He was plainly aware that in every man's passage through life there were times of overwhelming sorrow. But optimism was the center of his personality, and he had the ability to rekindle fires of hope in those who were caught in the grip of despair. His faith did not permit him to accept tragedy with dumb resignation or with cynical realism. Faith and hope lived together, and faith made hope strong—if necessary, raising it above reason. It was a gift to those active in their own cause. In desperate times, therefore, he could always turn to one special kind of activity—prayer, which he thought had the power to move the world and to achieve the impossible.

He disliked austerity. His strong appetites cried out and demanded to be filled. He saw nothing wrong with this, and he believed there was little merit in depriving himself of th55e comforts that had come to him. A poor man's life had few pleasures. If the nobleman won peace for a restless conscience by giving up things or by accepting self-inflicted pain, then let him, but one honest feast was worth a thousand fasts. Besides, the austere man too could be a monster of iniquity. The entire problem of punishment whether by God or by man did not interest him. He did not like to talk about the fires of hell or damnation of the wicked. He was more interested in original virtue than original sin. Suffering and sin were mostly the result of evil coming from ignorance rather than to a conscious will to transgress divine law. All men were potentially good, although they might do bad things. Gregory's faith was not a religion of wrath, nor was his God jealous nor punishing; He breathed love and talked of peace instead of retribution. He was the tsar of heaven, giving faith and consolation to His people—all of them.

§

Next, he turned to a problem attracting his deepest attention and causing him great concern: the linking of sin and salvation.

In seeking his path to salvation he was driven by a riddle: if in him flesh warred with spirit, how could he ever hope to find peace? He tried to abandon his worldly pursuits and live a life of holiness, but his backsliding always drove him into vice-ridden ways. He noted that his life adopted a kind of rhythm: moments of sin alternated with periods of repentance. Even the greatest efforts of will could not put an end to his trespasses. Could he hope to live the kind of life his desires dictated and at the same time follow a ministry dedicated to helping others? Was he not destined after all to be a man of God? Was perfection to be denied him because of human frailty? He wondered why God had placed two clashing desires in a man like him, a weak vessel addicted more than most men to sins of the flesh.

Here again the lives of the saints offered little guidance. While some of them mentioned carnal temptation, they kept a chaste silence, by tradition, on how they finally overcame it, except to mention an occasional self-burial or the wearing of chains. Clerics counseled him to purge himself of sin through prayer and asceticism. In keeping with this advice he learned to pray with concentration and fervor few could equal, dropping himself into a spell at a moment's notice and losing awareness of his surroundings, or passing long hours into the night on his knees in church. Sectarians might hold up to him the example of their hero, Avvakum, who after being tormented by desire, quieted his lust by lighting three candles and fitting them to a lectern; then he placed his right hand in the flame and held it there "until the evil passion was burned out." But such zealotry from a man who exhorted his own son to perish by fire was not for Gregory. The lesson here was plain: salvation was not to be found either in fanaticism or in asceticism. He suspected that the man who renounced normal desires was probably an incomplete being who, seeking to make a virtue of necessity, elevated his imperfections into virtues. No fires could destroy Gregory's passions, he had to live with them.

At first he regretted his periodic plunges back into sin, but even in moments of depravity he was gripped by a feeling that he had a mission, that he could achieve perfection. He refused to believe that the traditional concept of holiness was the only one, and he suspected that somewhere there was another kind, one that fitted him, that would not force him to make an agonizing choice between flesh and spirit.

Within the limits of traditional Orthodoxy there was no solution to the dilemma that Gregory posed for himself. Since he was unable to control his conduct he had to discover rational explanations for it, far-fetched as they might be. His behavior, which was partly beyond his control, was bound to shape his definition of a Man of God. He yearned to feel that by his own merits he had become a holy man and won a degree of perfection in this world. Most of his beliefs were forged in the struggle to make his thoughts conform to his actions, or to find a way in which his convictions could be brought into harmony with his behavior. He wanted to die and to be born again, to be transfigured and know without doubt that there was in him a spirit that gave him the power to do miraculous things. To a man with such strength would be given the ability to transmute sin and make it good. He claimed that if divine fire possessed him all evil would be burned out and therefore all his actions would be good. Sin was not attached to the act but to the doer. A man who did not have the inner grace that came with perfection would sin even in trivial ways; the perfect man could not sin even when submitting to the needs of the flesh. Gregory was not a Manichaean; flesh came not from the devil but from God and therefore its demands were good.

His reason led him on. The greatest evil was pride, the crime of Lucifer and Adam. No other transgression of God's law was equal in seriousness. Gregory agreed that the spirit must brand the flesh that it might live, but it was well to remember that a Nemesis lurked in the denial of the body, and if he would be perfect he had to strive to evade this danger lest in his vanity he run the risk of being destroyed. To be humble was the only answer. But if nature tended to make him proud, then he had to use strong weapons to purge himself of such feelings, make himself aware of what an abject sinner he was. He had to commit a great sin in order to remember the worse evil that hid in pride of righteousness. Later, his enemies accused him of taking this argument one step further and adding that there was one kind of sin that could save the sinner. They may have been correct, but the evidence is not complete. He is said to have argued as follows. In sexual sin he could debase himself until the soul plumbed the lower depths, but such a death of the soul gave the opportunity for regeneration. In other words, he had to fall in order to prepare to rise to a state of holiness.

Such doctrines were not unique, although some of his contemporaries regarded the amalgam of piety and sin as an outrageous concoction which only the depraved Rasputin, with characteristic effrontery, would dare present in the guise of religion. He seemed to be creating a monstrous perversion of an Orthodox tradition that if the soul was sanctified the body might be too. Throughout the history of Christendom, however, such ideas as Gregory entertained had been held by individual prophets (such as the seventeenth century mystic Miguel de Molinos) and sometimes by entire sects. Gregory probably never heard of these people. The resemblance in ideas comes from their sharing the same premises and facing the same problems. Versions of such ideas passed around among the villages of the Russian empire. In his novel, *Satan in Goray*, I.B. Singer tells of a religious adventurer who comes to a Polish town and preaches the doctrine of salvation through sin.

On the roots of Orthodoxy that still remained alive in him, Gregory grafted ideas plucked from his own fancies or from his observance of sectarians. Some of these groups, in order to solve the problem of achieving what they considered to be a sanctified marriage between a man and woman who could not accept the blessing of a state priest, experimented with different kinds of formal and informal unions. At the same time they wanted earnestly to preserve as much as possible of the trappings of traditional Christian beliefs. Gregory found among such people a richness and confusion of thought that provided him with some justification of his own faith. In this way he shaped his notions of abasement through sin as a means of salvation. Through this maze he consulted the inner voice that he regarded as an infallible guide. As time passed he became confused and thought that all inner urges, including the base ones, came from some divine source of wisdom that constantly revealed itself to him.

§

The years of youthful disillusionment were over; the time of fulfillment was about to begin. Fear and discouragement no longer pursued him. With cautious confidence he created a cry of his own, as the New Testament[13] put it; now the moment had come when he had to put his ministry to the test in order to see if it were real. According to traditional practices of the Wanderer, this test had to take place among people who knew him before he set out on his wanderings.

---

13. [Editor's note:] The twentieth century Knox translation of 1 Corinthians 1:12.

III

THE ROAD TO COURT

Gregory's public ministry began when he returned to Pokrovskoe after a period of wandering early in the century. It is likely he concluded that he had reached a stage of development where he might test his religious call before the people of the village. Their reaction would determine if he were a true Man of God or *starets*.[1]

To his neighbors he confided that he had learned the road to salvation and he could show others the way. In search of privacy, he conducted meetings, first in his house, then in the mews nearby, and finally in the woods. Such activities aroused fear and ire among some of the local people. They began to whisper about these gatherings, pointing out that they were composed mostly of women, they featured strange religious ideas, and seemed to substitute the rhythmic chants of the sectarians for the hymns of Orthodoxy. These rumors eventually reached the ears of the village priest, Fr. Peter, who began to watch Gregory closely. Since returning, the Wanderer had been a properly respectful member of the flock, but the persistent stories of his misconduct kept the fears of the priest alive. Such priests at this time were under constant pressure to be alert for indications of heterodox thinking among peasants. Alarming stories

---

1. Fülöp-Miller, pp. 33-51. Jean Jacoby, *Raspoutine* (Paris, 1935), p. 10. Alexandre Spiridovitch, *Les dernières anées de la cour de Tsarskoïe-Selo* (Paris, 1928), I, 117-23.

circulated about how swiftly sectarianism could fasten itself on a village and sweep away all of the faithful. Such a disaster could take away the livelihood of the priest. Even if the situation did not go that far he could be punished for lack of promptness in reporting dangers. The diocesan consistory might suspend him and thus deprive him of income. In this situation his family would be reduced to begging among the peasantry until he could gather a sufficient bribe to win reinstatement. Fr. Peter—whatever the extent of his objective suspicions—apparently decided to be prudent. He dispatched a warning to his superior, Bishop Anthony of Tobolsk, that the village seemed to be harboring a *khlyst* (Flagellant). Of all the sectarian cults that appeared among peasants, this one was most feared by the village clergy because little was known about it beyond its alleged ability to seduce a flock with its gospel of sex and emotional religion.

Church officials, as lowly clerics, were under much pressure to be alert for the threat of sectarianism. They were told by government officials, who linked religious and political rebellion, that the best time to strike such a danger was in the early stages, before it got out of hand. The formidable K.P. Pobedonostsev, Over Procurator of the Holy Synod, had sternly informed them they must be alert and he himself was active in this campaign against heresy. He visited "infected areas" as he called them, to dramatize his personal involvement and interest.

Almost at once a cleric, Fr. Berezkin, appeared in Pokrovskoe. He was a specialist in sectarianism and an inspector of the Tobolsk Ecclesiastical Academy. Experienced in such Investigations, he began to look over the evidence and to make attempts at gathering facts. He ordered the police to raid the Rasputin house several times in search of proof that the premises was being used for illegal meetings. No evidence was found. One of the police actually learned to like and admire the accused and greeted him as "Father Gregory." When these searches failed the case was dropped. No recommendation for prosecution was handed up to the district attorney, but a file describing the incident was kept in the diocesan offices in Tobolsk, and a copy was dispatched in routine fashion to the Holy Synod in St. Petersburg. And so began the charge of Flagellantism that dogged Gregory the rest of his life and pursued him even after death.[2]

---

2. Even at this early date his progress toward a career as a peasant Man of God was abnormal. His reception in Pokrovskoe was not a clear mandate for him. He was not universally acclaimed or advised to go to the capital. In St. Petersburg he was received in the same way; although he had a small band of devoted friends and sponsors many

§

In Pokrovskoe, investigations and the excitement surrounding them drew attention to him. He felt uncomfortable in the light, where all eyes could follow him; he instinctively preferred shadows. Only one path now seemed open to him: to go to St. Petersburg. This trip may have been a strategic retreat. but he was really following the tradition of the Wanderer, who usually capped his career with a visit to the shrines around the capital. There the Wanderer could evaluate himself by finding out if the church dignitaries recognized In him the spiritual qualities he believed he possessed.

On a number of occasions Gregory had visited Kazan, a large town on the middle Volga, and at the critical moment, after leaving his native village he decided to stop off there, a place where he could count on the help of friends.³ Kazan was the hub of much Russian academic and cultural activity in church affairs, and it was the training ground of workers destined to serve the missions of Russian Asia.⁴

In his many trips to the city he had made friends, such as the wealthy furrier, Katkov, who later went to St. Petersburg with him. A publicist, E.J. Dillon, who travelled as a journalist in Russia and who was a confidant of the Chairman of the Committee of Ministers, Sergei Witte, wrote that in Kazan and several other Volga towns he had met "converts"—probably "friends" would be better—of Rasputin.⁵ For Gregory, Kazan was really a jumping-off place, the spot where he was discovered by people who were important enough to be able to launch his career.

It may have been Katkov who brought him to the attention of the clergy in the Ecclesiastical Academy of Kazan, one of the leading church schools of higher learning. The academy was bustling with the activity of several prelates intimately connected with high ecclesiastical circles in St. Petersburg. Gregory scored a success at the school, where he ingratiated himself with the seminarians and their leader Fr. Michael, who later left the church to become a champion of the sectarians. Gregory also got the

---

people regarded him with suspicion and resentment.
[Editor's note: a colorful reimagining of this scene can be found in the Appendices.]

3. Rasputin, *My Father*, pp. 27-33
4. John Meyendorff, *The Orthodox Church* (New York: Pantheon Books, 1962), p. 119.
5. E.J. Dillon, *The Eclipse of Russia* (New York: George H. Doran Co., 1917), p.210.

support of the head of the academy, Bishop Andrei, and the curate, Chrisanthe, head of the Orthodox missions in Korea.[6] These men became his companions and sponsors. To the priests of the sister academy in St. Petersburg, the major monastery or Lavra of St. Alexander Nevsky, Andrei recommended him as a peasant *starets*. Gregory exercised some care in hiding the unorthodox side of his religious views, but word of his nightly sprees in the poorer quarters of the city began to spread. The priests of the school refused to listen to the tales. Toward the end of 1903 Bishop Andrei sent Gregory to the capital in the company of Chrisanthe.

Gregory's critics were not surprised that he got to court, but they were astonished by his ability to get his foot on the ladder of ascent in the church. His quick wits and good luck—combined with prudence and intelligence—helped him to make his way. But mostly he was the right man in the right place at the right time.

§

The leadership of Orthodoxy was concentrated in St. Petersburg, where all the complicated apparatus for controlling and directing the day-to-day operation of the church existed. But the routine problems of daily administration did not create the air of tension that was the background of Gregory's dramatic emergence. The capital was the one place where the leaders gathered to watch general trends and to guess about the future. As they looked at the future of the church they saw nothing but trouble. Because the atmosphere was like that of a command post, unclear threats, small alarms, and growing crises created a feeling of anxiety often bordering on panic, the kind of panic that forced men to dream of saving the day with quick-working remedies.

Some men responsible for the welfare of the church were deeply troubled about the trend of its recent history, and they were aware that in the future it was facing a voyage into stormy seas. The disasters that struck it in the twentieth century did not come as a surprise to them. But most clerics were unconcerned; they were confident that its destiny had been assured by God.

Among concerned leaders some called for reform. But before this could happen they had to leap over a hurdle: the government. Tsarism

---

6. Bienstock, p.100.

held the church in a grip of desperation and would not let go. In the past the government had strangled patriarchs and killed bishops, and it treated the lesser clergy with contempt. It had chosen the church as one of its main props and dared not relax its hold. Some clerics of St. Petersburg and Moscow believed that as the state sank deeper into the morass of political crisis it was dragging down the church with it.

Some thought they saw the answer to the church's problems in the calling of a great council (*sobor*) of clerics to discuss problems and to suggest answers.[7] There was much quiet support for the notion of letting the church go free provided it was not disestablished and could still count on the strong arm of the government in dealing with its competitors and detractors. The delights of freedom were evident to churchmen, but they still had not learned to live with freedom's risks, and so in a way they kept their own chains from falling off. For a long time rumors had been heard that the tsar was thinking about the calling of a *sobor*, but he had repeatedly failed to act. As time passed many bishops lost hope and complained bitterly that the government was deceiving them and could never be trusted.

Many of the leaders among the clergy were to be found in the St. Alexander Nevsky Monastery and its Ecclesiastical Academy, situated on the banks of the Neya River that flowed through the capital. Although the monastery had played an important role in ecclesiastical history, in the last quarter of the nineteenth century it had declined.[8] Tsar Alexander III on visiting it once was angered to find that the careless monks who were supposed to be in attendance on the art treasures and tombs were not on duty.[9] The clergy were, in fact, interested in other things. Bishop Sergei, a blonde, dreaming mystic later to be patriarch under the Communists, was concerned about folk religion and consulted often with Fr. Benjamin, an authority on peasant faith.[10] Fr. Medved was the confessor of the Grand Duchess Militsa and had contact with circles close to court. Theofan was a learned and ascetic teacher who after 1908 became rector of the academy and confessor of the empress. Fr. John of Kronstadt, the best known churchman in all Russia, often met and talked with the monks; Bishop Anthony of Volhynia, the most ambitious churchman in Russia, kept closely in touch with them by letters. There were many oth-

---

7. Curtiss, pp. 211-14

8. M.J. Rouët de Journal, *Monachisme et monastères russes* (Paris, 1952), p. 150

9. M.N. Pokrovsky (ed.), *K.P. Pobedonostsev i ego korrespondenty, pisma i zapiski* (Moscow, 1923), I, pt. 2, 643-44. "Kniazia tservki," *KA*, XL (1930), 102.

10. Spiridovitch, *Raspoutine*, pp. 42-44.

ers—teachers, writers, high-ranking administrators and men with chaplaincies in some of the most influential families in the realm.

The ferment in the academy begot meetings and discussions in which the leaders talked not only about church problems but also about the need for freedom in society as a precondition for freedom in the church. Among them were men who were the first to see that the threat of mass atheism was coming to Russia as it had to portions of western Europe, and they were beginning to suspect that the religious question was connected with social and economic questions.

There were those who hoped all problems could be solved by a more traditional means, a short-cut that could be opened by the right contacts at court. If the emperor could be won to their cause then their difficulties would evaporate. But since 1900 they had been watching the imperial family with growing apprehension: the tsar and tsaritsa were seeking comfort in strange religious experiences, including spiritualism, it was said. Alarming rumors had been heard that they were seeking solace in a parade of grotesque prophets, startsi, doctors of hypnotism, holy dwarfs, and crazy Wanderers—all of the people who had queer religious ideas.

Who were these mysterious people? What did they want? These were the questions on the lips of clerics and political figures who feared the spread of sectarian religion. A secret conference of churchmen met and expressed the fear that the court visitors "might easily have been sent for unknown purposes" and that "the sectarianism emanating from the court might spread among the upper circles of society and injure the church."[11] They agreed that there was a need to exercise a moderating influence on the religious inclinations of the royal pair.

Some of the defenders of the old regime viewed the visitors as agents of a conspiracy. Such churchmen drew close to conservative political figures who believed that throne and altar had to stand together. They took their ideas from a body of literature found in the French ultraconservative press, which claimed that all revolutionary movements in Europe were part of a plot to weaken and destroy established regimes everywhere. In the eyes of these thinkers the explanation seemed plausible. It accounted for the inability of tsarism—despite the Great Reforms of the 1860's and 1870's—to establish effective leadership in the struggle against forces attacking society from within. Subverters, they thought, aimed their first thrusts at the upper classes whose faith in the regime was undermined, and then they turned to the peak of the social order

---

11. Rodzianko, p. 3.

and tried to influence the sovereign. Some of the known transients at court (Dr. Gérard Encausse, Philippe Vachot, and A. A. Taneev, father of the woman who later became the only friend of the empress (were rumored to be Masons, and therefore, it was believed, might be involved in a conspiracy.

When they looked back over the frustrations of recent years churchmen thought they saw the source of their troubles: the decline of their influence at court. Officials of the church were still received at the palace, but only on a few state occasions. Nor were they welcomed into the bosom of the Imperial family as they once had been until the death of Alexander III. Easy access to the family—which permitted some of them almost unlimited freedom in the palace—was now a thing of the past. None could get close to the tsar. In the presence of the empress they sensed a coolness and haughty disapproval. They recalled that she was originally a Lutheran who before marriage claimed that she would suffer much anguish if she had to exchange her original faith for Orthodoxy. But then suddenly and without public explanation she changed—or had she? Furthermore, she was lately developing a great friendship for The Grand Duchess Militsa, a former Montenegrin princess, whose ties had been to the Serbian branch of Orthodoxy, and who now was reputed to be taken up with oriental religion to the point where she had been writing articles on eastern faiths. They feared that the impatience and indifference of the court today might tomorrow change to hostility, especially if poisonous words were being placed in the ears of the imperial couple. Finally, they were taken by surprise when an official announcement sponsored a plan to canonize St. Serafim of Sarov. Some supported the plan with the hope that it might win back the loyalty of the emperor. Serafim was a popular folk saint and *starets,* and they knew the court was interested in popular religion and holy men. But the role of the emperor in the drawn out and dispute-ridden canonization process only highlighted the scant respect for church law that he sometimes exhibited. Nicholas and Alexandra were not turned around; moreover, it was rumored that they were continuing in their old ways, fascinated by false prophets.

It was plain that some kind of more imaginative tactic would have to be used if the tsar were to be won over. Perhaps the church could place a man of its own near the emperor. But a basic difficulty remained: no one knew what was happening at court. It was a closed world into which no eyes could see clearly, but the facts about the recent favorites, Philippe and Encausse, caused deep fears to grow among all upholders of autocracy and Orthodoxy.

The most colorful visitor was Philippe, a protégé of the princesses, Militsa and Anastasia, daughters of the prince of Montenegro, a small Balkan kingdom.[12] Militsa was the wife of the Russian Grand Duke Nicholas Nikolaevich. Anastasia was married to, but living apart from, the Duke of Leuchtenberg. She was a close friend of Grand Duke Peter Nikolaevich. At the end of 1906 the Holy Synod granted her a divorce and in the spring of 1907 she tarried Grand Duke Peter. The dukes were brothers and they were also uncles of the tsar. In September, 1901 the imperial couple made their second state visit to France to celebrate the Franco-Russian Entente which by this time was ripening into a full alliance. The Grand Duchess Anastasia interrupted the busy schedule of Nicholas and Alexandra to introduce them to the outlandish faith-healer, Philippe, who had left behind an adventure-filled career in the city of Lyons where he was an exponent of his own version of Christianity. At a military review at Compiégne, the tsar and Philippe were observed discussing something earnestly while impatient dignitaries waited nearby.

The emperor's high regard for his new acquaintance was demonstrated in an incident happening shortly after the royal party returned home. Nicholas and Alexandra, along with Peter and Anastasia exerted much pressure on the French and Russian governments to give a medical degree to Philippe, who had once been a medical student.[13] At first the officials refused, but they were eager to avoid any unpleasantness lest the growing military convention between France and Russia, as part of the diplomatic arrangement binding them, might suffer a setback. Gen. A.N. Kuropatkin, the Minister of War, had a low estimate of the emperor's

---

12. Philippe Encausse, *Sciences occultes, Papus, sa vie, son oeuvre* (Paris, 1949). Louis Maniguet, "Contribution à l'etude l'influence des empiriques sur les malades. Etude médico-sociale. Un empirique lyonnais: Philippe, "*Faculté de Médicine* et de Pharmacie de Lyon. Anee scolaire 1912-1920 (1920). Joseph Schewaebel, "Un précurseur de Raspoutine, le mage Philippe," Mercure de France CXVII (1917), 637-47. Leon Weber-Bauler, *Philippe, guérisseur de Lyon à la cour de Nicholas II* (Boudry-Neuchatel, 1944). Gérard Encausse authored many works, but most of them are now hard to find. Among representative works of his in the Bibliotheque Nationale are: "De l'état des societes a l'époque de la révolution (1894), "Comment est constitué l'être humain" (1900), "L'âme humaine avant la maissance et après la mort" (1898), "Du traitement externe et psychique des maladies nerveuses" (1891), Catholicisme, satanisme et occultisme" (1897). All these were published in Paris. See also Maurice Paléologue, *An ambassador's Memoirs* (New York: George H. Doran Co., 1925), I 202-206; Maurice Bompard, *Mon ambassade en Russie, 1903-1908* (Paris, 1937), pp. 26-29; Henry Rollin, *L'apocalypse de nôtre temps* (Paris, 1939); S. Iu. Witte, *Vospominaiia* (Moscow, 1960), II, 262-63, 273-75.

13. Abel Combarieu, *Sept ans a l'Elysée avec le président Emile Loubet* (Paris, 1932), p. 167.

abilities and a fear that he might do something foolish when slighted. Consequently, the general made Philippe a sanitary inspector or military doctor.[14] The matter was settled.

The crude pressure tactics and unreasoning insistence that the degree be awarded to a man with a suspicious past and insufficient formal training left behind an unfavorable impression of the tsar. In addition, there was much concern about Philippe's mission and his connections. The few who saw him reported that he was a straight-forward man of simple tastes and charming personality whose unsophisticated speech revealed his humble origins. He freely admitted he was a Savoyard peasant. Calling himself a man of the people, he presented a political faith centered on the masses, whose loyalty and goodness he proclaimed to the emperor. The corridors and drawing rooms of the palaces at Tsarskoe Selo and St. Petersburg buzzed with resentful remarks about this parvenu who would not share with anyone his intimate talks with the royal pair. But he lived by his own gospel of forbearance. He refused to defend himself, saying that he loved his enemies and would silently and humbly accept their assaults.

Encausse, who used the name "Papus" among French spiritualists and occultists, went to Russia on trips between 1901 and 1906 where he made many friends in aristocratic circles. It is not known how close he was able to get to Nicholas and Alexandra, but they probably learned about his views indirectly. There was a belief in official circles that he was often consulted by the tsar, an idea that was certainly incorrect. Unlike Philippe, who was his friend, Encausse did not use faith alone for healing; he had taken a medical degree at the Sorbonne in 1891. However, the public knew him mostly as a prolific author of books on cabalistic lore.

Encausse's ideas resembled those of Philippe and Rasputin. For instance, he said that all faiths were one, that in the final judgment all sins would be forgiven except one—pride, and that suffering was not God's curse nor a sign of His displeasure. He was a spiritist rumored to be conducting table-thumping sessions at court. But spiritualism attracted the support or curiosity of a wide range of important people all over Europe toward the end of the century when it became a rage. To some it was a parlor game, to others a genuine way of talking to ghosts. Nicholas II indulged in meetings with spiritists, but it may have been only an amusement to him although there was a tale circulating that he had tried to raise his father's ghost at seances.

---

14. Bompard, p. 28.

The problem for most observers of these phenomena was to find out who were Encausse and Philippe and what they were trying to do. It was not easy to know how they were related, and as a result many observers have confused the two. Encausse modestly insisted that Philippe was his master and teacher. This may have been a wise tactic because the emperor and empress would be more impressed with ideas known to come from a former peasant. But whatever the relationship and whatever the goals of the two, the lesson was plain: to influence the Russian government get your man close to the tsar.

Among the clerics at the Lavra who watched all these events was Sergei Trufanov, called Iliodor, who took the habit at the end of 1903, graduated from the academy in 1905, and went on to play an important role in the early career of Gregory.[15] Trufanov shared the dreams and fears of many of the teachers and students at the school. He had been born a peasant, a Don Cossack, and had entered the academy full of zeal, with a desire to help the poor through work in the church. After becoming a monk he was introduced to the social world of the capital. One day a deacon took him to Strelna, a fashionable town twenty miles from St. Petersburg, and showed him a Street containing the villas of the mistresses of some of the aristocrats. Iliodor, an aggressive puritan, was outraged when he saw these and then thought of the millions of suffering poor who needed the money spent on the sinful pleasures of the rich.[16] He was deeply offended by what he saw among the nobility: their lack of concern for the people, their frivolousness and readiness to mock the church. "I saw the Arthuroviches and the Eduardoviches," he complained, noting that these simpering aristocrats spoke their native tongue with foreign accents, and sneered at tsar and peasant alike.[17]

He became active in reactionary movements that were supposed to save Russia. When the Black Hundreds were organized early in 1905 he quickly became a member and leader.[18] But he and Fr. Theofan (one of the leaders at the academy and a close friend of Iliodor) thought that the key to successful action was in the possibility of placing at court a peasant *starets* who could be counted on to represent the cause of Orthodoxy and autocracy. Only in this way could the power of the state be put at the disposal of reformist churchmen. For the moment, however, the wrong kind of people were being received at court.

---

15. Trufanoff (Iliodor), *Mad Monk*, pp. 19-21.
16. Trufanoff, *Rasputin*, p. 43.
17. Ibid, *Monk*, p.36
18. *Soyuz russkago naroda* (Moscow, 1929), p. 4.

At this moment, in the spring of 1903, word arrived that a *starets* was on his way, coming from the academy in Kazan, which recommended him highly. He had the blessings of Michael, Bishop of Kazan. When the monks talked to this *starets*, Gregory, they decided he was the man they had been praying for. Theofan planned to introduce him eventually at court. Past experience had taught that the empress was interested in peasant holy men. This one, who seemed to possess in abundance all the characteristics of the Man of God, might be able to replace the itinerant holy men in the palace. But first the plan had to be developed slowly and carefully and the ground had to be well prepared.

Gregory was in and out of the capital a number of times during the next two years. At first he stayed in the Lavra where he enjoyed the long discussions with the students who he impressed with his knowledge of Scriptures. Later he moved to the apartment of the economic writer and publicist, G.P. Sazonov, whose journal, *Russia* was sometimes critical of the government. Theofan then introduced him to some of the Important members of society, and he was soon invited to a number of influential salons. Many of the proprietresses of the salons, Countess Ignatiev, Baroness Korf, E.G. Shvarts, and others, were interested in religious questions, especially in the religion of the people.[19] Gregory became a permanent guest at the gatherings of Countess Ignatiev, whose worldly receptions on Sunday evenings were followed by ecclesiastical salons on Mondays that provided a meeting place for politicians and persons interested in promotions in the Holy Synod. It was said that all future Over Procurators of the Synod were to be found among her guests.

Gregory was careful to appear as a loyal son of the church. "In this milieu the *starets* naturally presented the modesty of a young virgin and the prudence of a serpent," as one police official put it.[20] A lone protest came from Bishop Anthony of Volhynia, who had heard of Gregory's activities in the dens of Kazan, but the bishop's warning was brushed aside by the excited clerics in St. Petersburg. Anthony, who saw himself as the future patriarch, was listening by means of an extensive and secret system of correspondents to all currents of opinion in church circles, and he was in a better position to know about the *starets* than were the bureaucrats and unworldly prelates in the Lavra. However, Gregory's motives were sufficiently mixed to confuse those around him, and there can be

---

19. Zhevakhov, I, 262. Anonymous, *Russian Court Memoirs* (New York: E.P. Dutton & Co., 1919), pp. 288-89.
20. Spiridovitch, *Raspoutine*, p. 61.

no doubt that he took his religious mission seriously and was able to convince others of his dedication.

After serving an apprenticeship on the high levels of society he got the feel of this new world, its ideas and prejudices, its standards of behavior, its opportunities and dangers. His eyes and ears took in everything, and he learned when to speak and when to be silent. While his friends thought they were using him and preparing him for the role they had designated, he was quietly getting ready for a much different kind of role, although even he was not aware of his destiny at this time.

After having lived in the capital for a while, he began to move about among different kinds of people: businessmen and elements of the petty bureaucracy. A typical associate of this era was V.O. Lokhtina, wife of a prominent engineer. She claimed that Gregory by his prayers had cured her of a disease the doctors could not cure. She at once became his devoted follower, introducing him to her circle of friends.

But he was never interested in developing a cult of his own. He had scored some impressive triumphs in the years since he began his climb, and had no doubt made some enemies too, but many challenges lay ahead before he could get to court, the goal of his sponsors at the Lavra. If he could, for instance, win the approval of Fr. John of Kronstadt he would have a victory worth almost any effort.[21] His friends could prepare the way for his presentation, but he had to rely on his own gifts to make a good impression on the archpriest.

No one who lived in Russia at that time was unaware of John. His face could be seen peering out with Christ-like serenity from inexpensive colored pictures hanging on the walls of most inns, post houses, and homes all over the empire. Although his influence at court had declined after 1894, he gained great fame as the pastor of Kronstadt, the large naval base on Kotlin Island in the Gulf of Finland, twenty-five miles from the capital. He was one of the rare members of the pastoral clergy able to capture the popular imagination and at the same time to win the support of the hierarchy. He was a confidant and confessor of many members of the social world. Constantly, he moved among them begging for money that he used in his parish for almsgiving, orphanages, inns, hospitals, and a large cathedral he was building.[22] His influence extended into many places. He was the idol of the restless young clerics interested in applying

---

21. A. Semenov, *Otets Ioann Kronshtadskii* (New York, 1955). Alexander Whyte. *Father John of the Greek Church* (London, 1898). John Sergieff (John of Kronstadt), *My Life in Christ* (London: Cassell & Co., 1897).

22. Juliette Adam, *Impressions françaises en Russie* (Paris, 1912), pp. 23-26.

the Gospel to social problems. In the cathedral there were always large crowds present whenever he celebrated the liturgy. On such occasions he was known to have visions, one time crying out in a terrified voice that he was seeing a vision of Golgotha.

John was a proponent of simple, traditionalist viewpoints. He disliked sectarians, Jews, and revolutionaries and thought the tsar should deal sternly with them. His theological convictions were correct and conventional, and many of them were the opposite of what Gregory thought. Nevertheless, Gregory could find good things to say about him. John did not bury himself in a monastery, he actively helped the poor and never showed an interest in rank, rising no higher than the position of archpriest. His sole vice was forgivable: he liked ornamental crosses hung around his neck and he wore fine silk cassocks. To Gregory, these were the simple ostentations of the poor man.

The approval by John would be a milestone in his career. His friends at the Lavra knew the value of the priest's endorsement, and they most likely worked hard to acquaint John with the facts, as they knew them, about the amazing *starets* who had come to St. Petersburg. John was surely able to appreciate the advantages of having a *starets* at court who could carry the point of view of the clerics. What Gregory needed was legitimacy, the recognition of his position as a Man of God. The blessing of John could accomplish this for him. His help given now to Gregory might yield him some reward later from a grateful *starets*. A successful meeting was probably arranged in private between the two, but the public was led to believe they never had met. Then, according to Gregory's friends, there came an awesome moment at a crowded service in the cathedral when John publicly recognized Gregory. He called him forward from the rear of the cathedral, introduced him as a Man of God, and repeating the prediction of the hermit-monk Macarius of Verkhoture, proclaimed him a man with a great mission. Under these theatrical circumstances Gregory won the approval of the famous priest.

The day for which Gregory had been groomed arrived during the momentous events of the Russo-Japanese War and the Revolution of 1905. Philippe had compromised his position at court by dabbling in politics and making wrong predictions about things he did not understand. He seems to have let himself be drawn into the circle of people who wanted Russian power to play a dominant role in the Far East, and he made bad predictions about Russian victories. He supported the canonization of Saint Seraphim of Sarov and said—as many others did—that God would respond to this canonization by granting the royal family an heir to the throne. In the summer of 1904 the Tsarevich Alexis was born. But Philippe was ailing and he was mourning the death of his daughter. He left Russia saying he would never return. He died in July, 1905 in France. Nevertheless, Alix [the empress] always remembered him with affection and in her letters written during the first World War she reminded Nicholas of Philippe's words.[23] She was especially impressed by his predictions, made as he took his leave, that Russia would always have an autocrat and that another holy man would soon come to serve her and her husband. She took this promise seriously and waited for the coming of the new Man of God.

Amid the clamor of war and revolution few noted the departure of Philippe. Fr. Theofan, who knew something about events in the imperial palaces, took note of the minor event and decided that Gregory's hour had struck; now was the time to bring him to the attention of the imperial family.[24] This could be accomplished, he thought, with the help of the Montenegrins, who by the spring of 1905 were very close to the royal couple, and therefore they had the necessary influence to help Gregory gain *entrée* at court.[25] The princesses had in fact recently demonstrated their influence when Anastasia succeeded in having Pierre Gilliard, the teacher of her children, appointed to the important position of tutor in the imperial household. Also, Anastasia had a friend, Maria Vishniakova, appointed court nurse.[26]

Gregory's friends had marshalled considerable support on his behalf: the priests and students at the monastery and academy, the Montenegrins, dukes Peter and Nicholas Nikolaevich, the leaders of several of the important salons, and a number of influential church dignitaries. All had different motives for wanting to see their man get to court, but all had something to gain—so they thought—from his being there. His clerical patrons now realized that revolution was driving autocracy to the edge of an abyss and no one could be certain of what the tsar might do.

---

23. Alexander Fedorovna, *Pisma imperatritsy Alexandry Fedorovny k imperatoru Nikolaiu II* (Berlin, 1922), June 16, 1916. Hereafter for this source and for the letters of the emperor to the empress the first initials only will be used.

24. A. Shumigorskii, "Rasputin v sudbakh pravoslavnoi tservki," *Istoricheskii vestnik* (March-June, 1917), 632.

25. "Dnevnik A.A. Polovtseva," *KA*, III (1923), 89. N.M. Rudnev, "Pravda o tsarskoi semei i temnykh silakh," *Russkaia letopis*, II (1922), 41. As an official investigator for the Provisional Government, Rudnev went through the family correspondence at Tsarskoe Selo. Some of this material was later lost. He stated that it indicated clearly the Montenegrins were responsible for bringing Philippe, Rasputin, and others to court.

26. Spiridovitch, *Raspoutine*, 72.

More than ever it was now important to hurry and to bring the *starets* to court, where he might be able to bolster the will of the sovereign who had already shown signs of breaking. As rumors had it, he was considering granting a consultative assembly. The final steps were quickly taken to get Gregory to court. One of the priests of the Lavra, Fr. Yaroslav Medved, was the confessor of Militsa. On the suggestion of Theofan he had arranged for Gregory to meet the duchess in the summer of 1905. The priest continued to bring Gregory to the residence of the duchess at the Znamenka Palace at Peterhof, a ten minute drive from the imperial palace. Militsa was captivated by her guest and said she was willing to introduce him into the imperial household. Count Witte, who always remained on good terms with Gregory, was eventually asked to help in this endeavor. He had returned from his successful trip abroad in which he brilliantly conducted Russia's peace negotiations at Portsmouth, New Hampshire with the Japanese. His stock was never higher, he knew he was a power in the government, and was in a mood to use his strength. However, if he did intervene in the introduction of Gregory his role was a modest one.[27]

The summer of 1905 was a grim time. The year had opened with a grave shock—the surrender of the fortress and armies at Port Arthur, thought by most Russians to be impregnable. At the end of May the Japanese in the classical naval battle annihilated the Russian fleet at Tsushima; more defeats came during the summer as the Japanese continued to drive back the Russian forces in Manchuria. At the imperial palace pessimism reigned. The empress relied more than ever on the comforting words of her friend Militsa. Toward the end of October the Montenegrin sisters and the dukes Peter and Nicholas Nikolaevich were constantly at the palace, visiting it every day and even taking their meals there although the tsar and tsaritsa were accustomed to dining *en famille*. The diary of the emperor reveals no such intimacy was accorded to any other persons.[28]

At this time the emperor went through the greatest emotional crisis of his life; even the death of his father or the trials of exile in Siberia in 1918 did not cause him so much anguish. The situation at court was frenetic. Almost none of the advisers who had formerly talked about the use of force and the need of a firm policy to check the revolution now thought that any measures could stop the revolution. Two officials, Trepov and Durnovo, hard and unyielding conservatives, made pronouncements

---

27. P.D. Shcheglovitov (ed.), *Padenie tsarskago rezhima* (Moscow, 1924-27), Beletsky, III, 388. Hereafter cited as PTR. The name appearing before the number of the volume will be the name of the person who gave testimony.

28. *Dnevnik...Nikolaia II*, pp. 221-24, October 12-31, 1905.

filled with a spirit of pessimism. The desperate tsar used an old tactic that had helped him in the past: he fired a number of reactionary members of the government, especially in the police institutions, and replaced them with moderates. Pobedonostsev bid farewell to the autocracy and was replaced by Prince A. D. Obolensky, a member of the Council of State, a man with a distinguished record of service in high levels of the government. But within a few weeks he was saying that only drastic concessions such as a promise of universal suffrage could save Russia. The retreats of autocracy merely whetted the appetite of the revolution which now realized it had the government on the edge of defeat. Witte undertook frequent trips to the palace and presented the emperor with a clearly drawn picture of the alternatives open to him: either create a military dictatorship or grant sweeping reforms. On 28 October the emperor played his last card: he asked Witte to appoint a commission to study the entire political situation. This tactic had worked before, but Witte brusquely swept it away, saying that the time was past for such manoeuvres. The final hour had come and a decision had to be made at once.

At court there was no help for Nicholas. In fact, some of the members of the royal house were preparing for the inevitable change of sovereigns and the transition to a form of limited monarchy. In a letter to his mother, who discreetly remained outside the country during these terrible days, Nicholas wearily complained of "the mountain which rests on my shoulders."[29] When he turned to his wife she suggested in a Delphic way that perhaps firmness might be the wisest policy, adding that in the last resort conscience ought to be his guide. Even though he let her express her opinions, he was not accustomed to asking her for advice on important matters of state.

The Dowager Empress did not advise reform in so many words, but she pleaded with her son to look to her friend Witte for help.[30] Witte already had made clear what he wanted: an imperial manifesto in which Nicholas granted extensive concessions.[31] Even the emperor's friends joined in the demands for change. An old tippler, Admiral Nilov, who often followed the tsar around and was the nearest thing he had to a crony, hinted that the autocracy had to give in. From the Caucasus came the word of the viceroy, Vorontsev-Dashkov, a family friend, warning that

---

29. Nicholas II, *The Secret Letters of the Last Tsar* (Longman's Green Co., 1938), p. 14, October 5, 1905.

30. Ibid, p. 19, October 16, 1905.

31. "Manifest 17 Oktiabra," *KA* XI-XII (1925), 73. He said that the only alternative was a military dictatorship.

the bombs and riots would not stop until a moderate government was put in power. A formidable array of people pressed Nicholas to grant extensive reforms. Some even hinted that stubbornness now could only lead to abdication later.

In the last days of October revolutionary violence grew worse. At Peterhof on the Gulf of Finland the court lived in a state of siege. The emperor had moved there so that he could communicate with his ministers in the capital by sea; all trains on the main Warsaw and Moscow lines had stopped, and service on the Tsarkoe Selo line might be cut at any moment. Russia was in the grip of a general strike. Life in the country was brought to a dead halt, but the revolution went on with even greater violence. The government was staggering toward collapse. Across the bay from the imperial residence loomed the menacing guns of the Kronstadt fortresses, where rioting sailors at any moment might take over the big naval base.

With danger mounting on all sides Nicholas began to weaken. Later he described this moment in a letter to his mother:

> I have the feeling that a whole year has already passed since the last letter I wrote...One had the sensation such as that one has before a violent summer storm. Everyone's nerves were stretched to the breaking point; one could not certainly endure this state of things for long.[32]

But he waited until the last desperate instant. To Nicholas the autocracy was something he had to guard with his life; it would be a sin to sign it away. But he was all alone. When he looked about desperately for help there was none. Witte refused to change his opinion: vast concessions had to be given. The tsar weakly assented and the officials began to draw up a manifesto. It was completed at 3 A.M. on the night of 29-30 October. When Nicholas saw it he gave in. He wrote later, "after having prayed to God, I signed it. My dear Mama, you just cannot imagine how tormented I was before doing that!" A valet noted that before signing he hesitated and made the sign of the cross.[33] In this way the autocracy disappeared forever. Russia was a constitutional monarchy. Nicholas himself sadly admitted this a few weeks later.

---

32. Nicholas II, *Secret Letters*, p. 22, October 19, 1905.
33. Alexis Volkov, *Souvenirs d'Alexis Volkov* (Paris, 1928), p. 57.

To the frightened advisers at court it seemed that the signing of the October Manifesto at first produced no results. There was a revolt in the garrison at Kronstadt, and at the naval base of Sevastopol in the Crimea rebellious troops seized the town. The same was done in Vladivostock, at the other end of the empire. The granting of political amnesty on November 4, as a further concession by the emperor did not bring calm to the capital.[34]

The failure of the manifesto to yield political gains in the first days of November threw the court into panic. Nicholas was sorely in need of soothing words to help sustain his courage. For a moment there was talk of even more concessions being necessary if the monarchy were to be saved. Nicholas must have realized that he might be spending his last hours on the throne. He was then told about the holy man in St. Petersburg and the Montenegrins volunteered to bring him to the palace for an interview. The emperor agreed. A short talk with the Man of God had the desired results. Nicholas was impressed enough to mention the meeting in his diary, although he often failed to note visits of ministers and important statesmen. At other difficult times Gregory would be called in to play this same role. After the defeats in East Prussia in 1914, Nicholas wrote, "This evening alone, under the peaceful influence of a talk with Gregory, my soul regained its poise."[35] In mid-November, 1905, the imperial government seemed to find courage. At conferences it decided that there would be no more concessions and no overt attack on the revolution—the government felt it had the moral force to make a firm stand and would permit the revolutionary gales to blow themselves out. The tactic was correct: within a month the revolution made its last desperate assaults. Peace soon began to settle on Russia.

---

34. Miliukov, *Historie de Russie*, III, 1113.
35. Nicolas (Nicholas) II, *Journal intime de Nicolas II* (Paris, 1934), p. 44, October 17, 1914.

IV

NICHOLAS II:
"What am I to do?"

Nicholas was born in 1868, the eldest child of Tsar Alexander III. His father was a placid man, but sometimes he rose to fits of Jovian rage, especially when he was asked to grant reforms. He presented the country with the awesome spectacle of an all-powerful tsar, a man of courage who feared neither foreign nor domestic foes. With his family he was a firm paterfamilias, but his sternness was balanced by his spirited wife, Maria Fedorovna, a Danish princess, whose sister was the wife of the future Edward VII of Britain. Under the stern eye of his Victorian papa the young Tsarevich Nicholas grew up to be a pleasant and unassuming youth, so unobtrusive that he was several times ignored at small receptions in the palace. He showed no unusual traits of personality, either good or bad, but was slow to mature. To Nicholas, his father represented the world of adulthood that he respected but feared. He thought his father would live for several score years, so he enjoyed his endless youth. But suddenly the tsar sickened and died, leaving the bewildered stripling on the throne.

Although Nicholas was reared to be tsar he was totally unprepared by education and personality for the job ahead. Like most princes of his time he was poorly schooled. A series of tutors worked intensively with him. They were picked for their loyalty to the regime and for the orthodoxy of their ideas; no one was much concerned about their teaching skills. Pobedonostsev, a senior tutor, was never able to get close to him. Nevertheless, the Over Procurator lectured him at length on the responsibilities of a tsar and also provided him with long disquisitions on state law.[1]

His academic coaching and the jumble of ideas thrown at him by various courtiers, generals, and palace servants prepared him inadequately for the role of twentieth century sovereign. Using standards known only to himself, he picked and chose from among the proferred ideas and accepted only those that appealed to his fancy. The result was a hodgepodge, a mind filled with curious contradictions and half-formed notions.

But he enjoyed the moments of lighter learning in the company of other young men who were his fellow officers of the guards regiment in which he served. Although the life of the guards was sometimes pointless and occasionally dissipated, it brought him closer to the world than he had ever been before. It also introduced him to his second brush with life, Thilde Kshesinkaia, who was about to begin a brilliant career when he first met her in 1890.[2] In the next year they started a flirtation that Alexander III ended with a conventional ruse: he sent his smitten son on a trip around the world, ostensibly for reasons of state.

In the summer of 1891 the acquaintance was renewed and it developed into the usual kind of liaison between a royal youth and ballerina. Before obligations of state took over his life, he enjoyed the company of Mathilde in a passionate interlude that was the final act of his youth.

Then the affair suddenly ended. The tsar's health failed quickly. Tradition was against a new tsar coming to the throne unmarried; the ceremony of coronation almost demanded a pair of rulers. Besides, he was of age now. He made his choice: it would be Princess Alix of Hesse-Darmstadt, a granddaughter of Victoria of England. From his parents and friends came a stream of objections: she was a dull child; there had already been too many German princesses who became tsaritsas and society was tired of them; there were better candidates available, including a beautiful French princess who would be a sensational catch—and a boost to Russia's foreign policy interests. The tsarevich had shown some stubbornness in refusing to put away Kshesinskaia and there had been

---

1. Pokrovsky, *Pobedonostsev*, III, 560
2. Matilda Kshesinskaia, *Dancing in St. Petersburg* (Garden City: Doubleday & Co., 1961) pp. 29-53.

family fights over the matter. Now for the first time in his life he made a stand. He was determined to marry Alix because he was in love. Pleadings could not move him, even when he was warned that for the good of the dynasty she could not be chosen. But the sad parents eventually gave in. Important events then came in rapid-fire succession in Nicholas' life: engagement, weddings, death, funeral, coronation.

But he was still unready to step into the shoes of Alexander III. It was not surprising that on taking the throne he turned to the Grand Duke Alexander and begged, "What Am I to Do?"[3] For him this was a question to be asked over and over again during his reign, and it was to be followed by another after he had lost the throne, when he considered all he had done: "Can I have been mistaken?"[4] History finally answered the second question. Nicholas kept on repeating the first, but he was never sure of any answer he received.

He faced an immense problem: how to save a Russia that was in deep trouble. Although he lacked an outstanding intelligence, he had enough sense to grasp (up to a point) the extent of the dangers hanging before his eyes. He was less ignorant of the dangers of Rasputin's presence than many people thought, but he could not act. The confused torrent of his emotions swamped his intelligence and his will. To solve such problems as the Rasputin affair demanded a ruler who would put great power at the service of informed opinion. Nicholas was not this kind of sovereign; he was not an autocrat at heart. He could be morosely stubborn, willful, and brimming with spite and persistence, but these were not the qualities of an autocrat.

His contemporaries bluntly expressed their feelings about the deficiencies of his character. The tsar had a weak will, they said. This refrain was often heard in government and court circles. It was repeated when people felt that he had not listened to their advice or he had flitted about and accepted all kinds of conflicting opinions. They were quick to note how different this procedure was from the practice of Alexander III, and they also saw the disastrous results it might have. Their experiences in dealing with Nicholas produced a uniform impression among them. As early as 1896 a prominent diarist, Gen. A.V. Bogdanovich, noted that at court a poem was circulating that commented on the power of shadows and the shadows of power.[5] Nicholas's mother, Dowager Empress Maria

---

3. Grand Duke Alexander Mikhailovich, *Once a Grand Duke*, (Garden City: Garden City Publishing Co., 1932), p. 168.
4. Rodzianko, p. 254.
5. A.V. Bogdanovich, *Tri poslednikh samoderzhitsa. Dnevik*. p.8, February 5, 1896.

Fedorovna, admitted to Witte that her oldest son had neither character nor will,[6] and Pobedonostsev made the same remark in 1900. D.S. Sipiagin, a devoted monarchist who died of an assassin's bullets while serving as Minister of Interior, complained that Nicholas was not to be trusted, that he was untruthful and deceitful.[7]

Nicholas knew what they were thinking, and the knowledge stung him. But he was determined to find the strength that would permit him to reign and to safeguard the autocracy. He could not do this by sharing the burden of rule; only he could exercise the autocratic power, for God had given it to him and he could not entrust part of it to any mortal—not even his wife. There were brief moments in the crises of 1905 and 1916 when he wavered in this conviction but then he quickly returned to it.

The problem then was not to worry about the opinions of society but to maintain the steadfastness of his will to reign. There were sources from which he sought to derive this strength. At the coronation in June, 1895 and again during the ceremonies at the start of the war in August, 1914 he walked along through the great crowds near the inside of the Kremlin. In Sarov in 1903 he moved through the huge gathering of half a million peasants who had come from all parts of the empire to witness the ceremonies of canonization of St. Serafim.[8] For years the police had warned him that presenting himself before such a mob would result in his assassination. Von Launitz, the governor of St. Petersburg, and Count Fredericks, Minister of the Court, were jostled but ignored. Even police officers in the imperial suite were not molested although these same men would not have dared to wander unattended on streets of the capital or Moscow. Such incidents provided the emperor with some idea of how the people felt about him. What he saw convinced him that they revered him. This awareness gave him comfort and strength, and no pleadings or warnings from officials could shake his confidence in the idea that no matter what happened the people would always be loyal.

The successful management of the state could not be based on the devotion of the peasant masses. Nicholas had to rely on groups that tra-

---

6. I. Vasilevskii, *Graf Vitte i ego memuary*, (Berlin, 1922), p. 12.

7. Viktor Chernov, *Rozhdenie revoliutsionnoi Rossii* (Paris, 1934), 97. Rasputin's estimate of Nicholas is interesting: "Something is missing inside." (*Y nego vnutri ne dostaet*).

8. *Sviateishii pravitelstvuiushchii sinod. Vsepoddaneishii otchet ober-prokurora sviateishago sinod po vedomostvu pravoslavnago ispovedaniia za 1903-1904 gody*. (St. Petersburg, 1909), pp. 1-13. A.A. Mosolov, *At the Court of the Last Tsar* (London: Methuen, 1935), pp. 133-35. Nicholas's fascination with Serafim caused much unfavorable comment among the upper classes. See "Nikolai II i samoderzhavie v 1903," *Byloe* (1918, no. 2), 201-02.

ditionally provided servants to fill the ranks of the administration. But these, the bureaucrats and nobles, were being alienated by many things. They resented the weak and ineffective leadership of the tsar, and his aloofness and secretiveness. In addition, all officials faced the possibility of death by assassination. Many began to wonder if the cause—and the leader—were worthy of such risk.

The hazards of service and the lack of recognition from the emperor naturally had their effect on the program of recruiting high-level statesmen. Among them there was a growing reluctance to serve. Therefore, Nicholas had to take desperate steps to obtain suitable ministers, and sometimes he was driven to order persons to take posts.

The nobility was the second group, after the bureaucracy, that was supposed to support the throne. But behind the facade of harmony between noble and tsar was a chasm of differences. The typical noble was resentful because he had gained nothing from the Revolution of 1905 although he had supported the tsar—out of desperation, of course.[9]

§

After the Revolution of 1905 all Nicholas's political problems settled around the matter of living with the new legislature, the Duma. He was not quite certain about what kind of government he was presiding over; he knew that the October Manifesto conceded a constitution, but he was not sure of how much power he had really lost. Was the legislature a competitor or a helpmate? His attitude toward it changed with circumstances. When it first met it trumpeted revolution; Nicholas hated it.[10] Next, he decided to give the Second Duma a chance and hailed its deputies as "representatives of the people." But he always felt uncomfortable while the Duma was in session. In October, 1913, he speculated, perhaps under pressure of reactionaries, about reducing it to a consultative position, but he went about this project in a scrupulously legal manner and abided by the decision of his ministers when they rejected his proposal.[11] While he bitterly resented the deputies and their words, his daydreams about silencing them did not include a *coup d'état*, and he once assured

---

9. P.N. Miliukov, *Rossiia na perelome* (Paris, 1927), p. 5.
10. *PTR*, Golovin, V, 372-73.
11. V.P. Semennikov, *Monarkhiia pered krusheniem. Iz bumagi Nikolaia II* (Moscow, 1927), pp. 92, 94-95. *PTR*, Maklakov, V, 195 *et seq.*

a British visitor that the Duma was a permanent feature of the Russian pioliltical scene.[12]

However, a continuing current of tension flowed between Nicholas and nearly all the deputies. Moderates suspected he harbored plans about unilaterally changing the electoral laws. As a result of these apprehensions the years of peace from 1907 to 1914 were filled with alarms and crises that could be brought on by almost any trifling word or rumor reported from the palace.

In the political struggle there were moments when the opposition hoped that the emperor might be educated to act like a constitutional ruler, but they wondered if he could be moved by arguments short of revolution. The supporters of autocracy, on the other hand, were afraid that the suspicion of the progressives—that the tsar could be influenced by his enemies—might be true.

§

Probably more than any other European ruler of the twentieth century Nicholas believed in the religious nature of the monarchy. The mystique of monarchy invested him, and his thought move in a world resembling that of Richard II. He saw that the coronation was more a religious ceremony than a political one.[13] It reminded him of the sacerdotal nature of his office. On him was bestowed the power for life and he was not permitted to rend the fabric of this power or to change it in any way. It was his duty to hand it on undiminished to his son.

Indeed, Nicholas warned the First Duma that he intended to pass on "a solid heritage." Because he weakened at a critical moment in 1905 and gave away some of his power, Russia suffered, so he believed. His warning, nevertheless, was not repeated in his last Duma address of 1916. At this time he realized that the constitutional question could not be postponed much longer. He probably hoped to put the whole matter on the post-war agenda of politics. Frequently, he made assertions about the future greatness of Russia, but he did not include himself in this picture.

---

12. Bernard Pares, *The Fall of the Russian Monarchy* (New York: Vintage Books, 1961), p. 118.

13. S.S. Oldenburg, *Tsarstvovanie imperatora Nikolaia II-go* (Belgrade, 1939), pp. 59-61. G.P. Georgievskii, *Koronovanie russkikh gosudarei. Istoricheskie ocherk* (Moscow, 1896), pp. 3-8, 117-19.

He refused to listen to demands of one of the dukes that the tsar appoint him to begin a study of the problems they would face at the peace conference after the war. Nicholas may have suspected that his own appointee would not be there. He never mentioned the post-war era except to say that the solution to many of the political problems pressing the country during the war would have to await the arrival of peace. He hopefully looked for his *Nunc Dimittis*. He had given his oath not to sacrifice the autocracy, and he probably thought that he could hand over the control of the state to the young Alexis, whose regent could then frame a coronation oath in which the new tsar would promise to cooperate with the elected representatives of the people.

§

This then is the picture of Nicholas II and his world. He saw himself as a ruler inadequately prepared for his role. He often resented helpful suggestions because he distrusted his ability to choose the right course in affairs, and he feared that any advice might represent a dilution of his autocratic powers. He was keenly aware of his own deficiencies and yet was crushed by a sense of duty that called him to rule in spite of them. Although he had not much faith in himself, he was sure that autocracy was good for Russia, so he determined to hold on despite his lack of ability. His ministers had little confidence in him as a man, and they had no respect for him as a ruler; they tried to perform their roles under him, but they quickly became dispirited. Since he was forced to deceive and reject them, they at first were resentful and distrustful, but soon their bewilderment changed to animosity. Dissatisfaction spread through entire groups of his traditional supporters.

With faith in divine assistance and the support of the common people he hoped that his reign would be a success. In a tsar, faith was more important than wisdom. He was sure that Russia's problems were no more complicated than this. It was hard to believe that anything was essentially wrong with the country. Although it was sinking deeper into political crisis it was prospering economically. Concessions had failed to put an end to the crisis; consequently, on the eve of the war he was prepared to try to return cautiously to some of the features of the old regime. During the years of the war he relived the era from 1903 to 1905 several times over as he swung back and forth between firmness and concessions. Alexandra was always at his side, ready to chide him about the first tactic and to cheer him about the second.

V

ALEXANDRA: LADY MACBETH

Princess Alix of Hesse-Darmstadt startled the royalty of Europe when she announced that for religious reasons she had refused to become affianced to the Russian crown prince, Nicholas. For generations German princesses had gladly left their poor and poky courts for marriages in the family of the tsars. They changed religions because they felt a Russian marriage was worth a liturgy.

She was fiercely pious and sentimental and although her faith was the center of her most passionate interests, it must be called religiosity rather than religion. She was one of those people whose creed made them turn away from humility: it girded her with righteousness and a spirit of smug assurance. Not for a moment did she doubt that she was more capable than most clerics in telling right from wrong, and she cherished a feeling of moral superiority over most other mortals.[1] Even Nicholas was not spared her chastening pride. During their courtship he confessed to her about his recent liaison with the ballerina, Mathilde Kshesinskaia. Instead of being shocked, or at least being discreetly silent as a good Victorian maid should have been, she launched glowingly into a lecture about how some people did not have the strength to resist temptation, implying

---

1. V. Poliakoff, *The Tragic Bride* (New York: Appleton & Co., 1928), p.115.

that they had to pay for their sins and learn how to lean on others, persons of superior moral qualities. She seemed almost happy to hear that he had once fallen.[2]

Charity and forgiveness were not in her. If a person conceded her excellence, she was capable of reacting with great tenderness. But she could be vengeful and unrelenting in her hatred, a vice that became clearly noticeable when she dabbled in political vendettas.[3] In these, her negative opinions were never tinged by compassion, no matter what the circumstances. Informed that Stolypin had just been murdered, she replied brutally that he deserved this end because he had overshadowed the tsar and taken on airs. But with Rasputin she was forgiving and understanding. When told that he molested women, she replied that the dirty-minded society of St. Petersburg thought only of carnality. Did they not know that the Apostles embraced and kissed? The critics could not understand the spontaneous affection of the peasant *abrazo*. If he sometimes got drunk it was because they seduced him with liqour, and if he had sinned once, well...he admitted that. Besides, her own husband had been a sinner. Did that mean he was always to be considered bad?

For her, religion was not supposed to win popularity; it was in part a consolation, a way that helped one endure the hardships of life, an assurance that one could live in righteousness alone when all the world was hostile. Although often depressed and melancholy, she was not a quietist. She could not be satisfied with the kind of vicarious experience that satisfied most of the faithful, but she had to participate in or be close to a direct experience. Mere contemplation was not enough. Prayer was the only kind of traditional religious exercise that interested her and she believed it gave her a direct approach to the living forces of faith. It was a political weapon more potent than ballots, or properties, or compromises. She did not pretend that she had a special power in prayer. Gregory, in her estimation had this gift, and his ability in passionate and concentrated prayer was second to none. She reasoned from this fact, and others she later discovered, that he must be a Man of God, and therefore through him she could exploit those parts of religion that interested her most: action and power.

She was too haughty to indulge in controversy. About her personal views on religion she was close-mouthed and made few attempts to proselytize those around her, especially if she suspected they might disagree.

---

2. *Dnevnik...Nikolaia II*, p. 69, July 8, 1894.
3. *PTR*, Guchkov, IV, 252.

As a result of her attitude little is known about the sources of her ideas or the way in which she arrived at them. Observers tried to define her belief by seeing only its external and obvious signs. The conclusions reached by this method were only in part correct, for she had a more systematic outlook than her critics realized. Religion was not only the overriding interest of her life but also the key to knowledge that helped her grasp the meaning of politics. An attempt to understand the sources of her religious views assists us in comprehending the motives and meanings behind some of her political acts.

General Alexander Spiridovich, who was in charge of the personal security police of the emperor, once mentioned a French book that was a favorite of the empress: *Les Amis de Dieu*. He did not see the book, he did not give the name of the author, and he was familiar with its contents only in a general way. Anna Vyrubova, the friend and attendant of Alix, said that the empress in 1905 gave her the book to read, but she was unable to understand it. Both the general and the lady-in-waiting stated that it contained the ideas of the Friends of God, a fourteenth century European religious movement.[4]

The Friends of God stressed the pietistic and emotional content of religion, and they drew attention to two things: the role of suffering and the need for humility, matters that interested Alix. Suffering was not pain without purpose; it was caused, the Friends said, by sin and it would have an end. Pain was not a sign that God had forsaken the sufferer, but an indication of His favor, even if it was inflicted unto death. Many of the Friends believed that the man who sinned was more likely to save himself than the man who thought he did not sin. Pride was the greatest danger to the soul. Because this was the sin they were most prone to, the rich and the mighty had to be unusually vigilant. Because they feared pride, the Friends rejected asceticism of almost all kinds, believing that it might make a person vain about his virtue.

---

4. Spiridovitch, *Raspoutine*, pp. 56-57. Anna Vyrubova, *Memories of the Russian Court* (New York: Macmillan Co., 1923). A to N, September 9, 15, 1915; September 7, 1916; June 10, 1916; October 12, 1916. For the Friends of God see: Sophie Buxhoeveden, *The Life and Tragedy of Alexandra Fedorovna, Empress of Russia* (New York: Longman's, Green & Co., 1929), pp. 133-34, 136. James M. Clark, *The Great German Mystics: Eckhart, Tauler, Suso* (Oxford: Basil Blackwell, 1949), pp. 5, 26, 69, 80-96. H. Delacroix, *Essai sur le mysticisme spéculatif en Allemagne au quatorzieme siècle* (Paris, 1900), pp. 36, 55-57. Encausse, *Sciences occultes*, p. 7. Xavier Hornstein, *Les grands mystiques allemands du XIVe siecle* (Lucerne, 1922), p. 216. Auguste Jundt, *Les Amis de Dieu au quatorzième siècle* (Paris, 1879), p.39. Anna Seesholtz, *Friends of God, Practical Mystics of the Fourteenth Century* (New York: Columbia University Press, 1934), pp. viii, 35-87.

The Friends were critical of the upper classes, saying they suffered from an excess of pride, and a lack of charity. True virtue and piety were to be found among the common people; to be poor was to be just. The true Friends, it was said, were poor in material possessions, and they might even be found among the most abject and illiterate portions of the population.

But a Friend was not easy to find. The great Medieval mystic and a leader of the cult, Tauler, said that most men were God's enemies; but like gold, Friends were hidden in the earth's unlikely places and could be discovered only by those who earnestly sought them. One of the few tests the seeker could trust was to note that the Friend was reviled by unjust men. Queens Elizabeth of Hungary (an ancestor of Alix) and Agnes of Bohemia said that there were Friends sent especially to help rulers.

Alix subscribed to many of these ideas. Her letters to her husband make it evident that she thought Gregory was a Friend who brought truth and wisdom. "Hearken unto our Friend," she said,

> believe in him, he has your interests and Russia's at heart— it is not for nothing God has sent him to us—only we must pay more attention to what he says, his words are not lightly spoken, and the gravity of having not only his prayers but his advice is great.

Later she said, "You remember, dans *Les Amis de Dieu* it says a country cannot be lost whose sovereign is guided by a man of God. Oh, let him guide you more." In September, 1916 she instructed Nicholas again, "Do listen to him who only wants your good...His love for you and Russia is so intense and God has sent him to be your help and guide and he prays so hard for you."

In her mind Gregory was cast in the role of the Friend. His own religious ideas, worked out during his apprenticeship, coincided with those she had accepted from the Friends of God and the talks with Philippe. When Gregory appeared in the palace her mind was half made up as to what he was and what he represented. However, she cautiously observed him for many years before letting herself be completely convinced that he was a Man of God. Gregory shrewdly sensed what was happening: he played his part with care and remained in character. And so his actions convinced her he was the man. One of most poignant features of the drama of their relationship was the coincidence that she was completely prepared by her readings and her religious conceptions to see him as the

Holy Man he wanted himself to appear to be.

Little interpolation is needed to point out the striking parallels between the ideas of the Friends of God and and the notions of Russian popular religious feelings. The piety of the common people all over Europe from Iceland to Siberia was similar until a generation or two after the Reformation. The source of Gregory's outlook, Russian popular piety, and the fourteenth century book treasured by the empress shared the same origins. Alix was not aware of the common historical connection of her ideas with those of Gregory, and for this reason the closeness of his and her views seemed to be providential incident.

To her there was no legitimate distinction between religion and politics. She regarded religion as a guide to political action. Fortified with her insights she came to think Gregory was something more than a human adviser; his words were more than common wisdom backed by common understanding. He had the key to all her problems and to Russia's problems as well. His words would eventually show her how to save the autocracy: he would help her to find the men who could be trusted to hold power. She would pass on these thoughts to her husband and insist he accept them. Thus Nicholas would indirectly receive the benefit of divine wisdom. God would see Russia's plight and the goodness of her sovereigns and He would save both the dynasty and the country. She did not believe that the life of the sickly heir was being preserved so that he might grow to be a monarch without a throne.[5] His life was a pledge to the future of the royal house. If she and her husband were true to the cause of autocracy, she thought, Alexis would live. It was not mere political victory or defeat that was at stake; it was the life of her son.

Whatever they might say about her in the capital, she was sure that the masses of the people revered her.[6] But she knew that a gulf separated her from them, although it was not as deep or as wide as the one which separated her from the upper classes. She felt that somehow the masses were close to her because they shared a common faith with her. She believed, as Nicholas did, in the cult of the people. This cult was plain to see at court where many peasant-priests reached high positions.

If she could not approach the people directly she would use an intermediary, Gregory. He soon realized what was expected of him, and he sought to convince her that not only was he a man of the people but also

---

5. *PTR*, Dzhunkovsky, IV, 501.

6. Wladimir Gourko (Gurko), "Le empereur Nicolas II et l'imperatrice Alexandra Fédorowna I," *La revue universelle*, XLIX no. 5 (June 1, 1932), 585, 587. Nicolas Sokoloff, *Enquête judiciare sur l'assassinat de famille imperiale russe* (Paris, 1924), p. 91.

he was worshipped by them. However, he had no popular following; all his instincts told him to avoid the limelight, to stay in the shadows. Later his friend, the monk Iliodor, provided him with a ready-made following.

But there was no intermediary to bring Alix close to the society—nor did she want one. The more she imagined the people loved her, the more the tragic apartness developed between her and the upper classes.[7] This alienation had begun before the coronation and continued until the end.

She was never able to establish her influence and importance in court and society. A tsaritsa was not necessarily the second most important person in the realm, for it was not the custom to ascribe a place of precedence to her merely because she was the wife of the reigning monarch. Other women of the imperial household, if they had the talent and personality suitable for the role, might be able to hold on to the position of leadership in the court.

Traditionally, Alix should have spent several years in transition from the position of wife of the heir to that of empress. As the wife of the tsarevich, for several years she could have learned the complex protocol and tradition of society and court. Instead she, a girl, a foreigner who was resented even before she arrived in Russia, was thrown into a situation that would have taxed the skill of a mature and sophisticated princess. On her first trip to Russia she had been hastily looked over by the imperial family and pronounced unfit to be the spouse of Nicholas. As she went back to Germany a failure, she could hear the mocking laughter of the court. When she returned a few years later in triumph she could feel the resentment. She knew society and court thought of her husband as a cipher, and she heard rumors that he might be replaced on the throne by his younger brother, the Grand Duke Michael. All these things made her retreat into a shell of aloofness and suspicion.

If her first child born in 1895 had been a son, Alix's position would have been much strengthened. As the mother of the heir she would have been entitled to the place and respect that the position of wife did not by itself afford her. But she reigned as empress for ten years without providing an heir, and as each birth was followed by increasing bad health, it seemed less likely that the succession to the throne would pass through the line of Nicholas. She declined in importance at court. As her health deteriorated during the first years of the new century, people began to suspect that there never would be an heir. Under these conditions the

---

7. Lili Dehn, *The Real Tsaritsa* (London: Thornton Butterworth Ltd., 1922), p. 59. Vyrubova, p.11.

Dowager empress remained active; she was the most important leader in society.

Tsarevich Alexis was born in the summer of 1904. Here was, indeed, a pledge from God that He would save Russia. Alix's faith had won a glorious victory. Now let her foes tremble! For a few days all troubles seemed to have ended for the imperial couple. But happiness was quickly snatched away from them when they made a terrible discovery: the boy suffered from hemophilia, an inherited disease that had killed her brother and other males in the line of Queen Victoria.[8] To friends and courtiers all seemed well and normal in the Alexander Palace, but the entries in Nicholas's diary reveal the kind of terror this discovery placed in the hearts of the parents.[9]

It had other consequences. Between them it set up a complex web of relationships. Alix might have been oppressed by feelings of guilt for what she had done to the child; after all, she passed on to him the disease that gave him no chance for a normal existence and at best gave him less than half a chance at life. But she transferred her burden to Nicholas, and of the two he was the only one who showed signs of being ridden by guilt. We can only guess how she accomplished this, and most likely she herself was not aware of how it was done.

As the *valide*,[10] the mother of the royal house and protectress of the dynasty, she had the responsibility of preserving Alexis's right to the throne. She believed that she had the will to carry on the fight. But always she felt the dread of impending tragedy. It stalked and haunted her because it could strike any moment. The child might trip or stumble, things a normal child did a dozen times in a day. A simple blow or any object falling on his hand or foot might end his life after days or weeks of agonized suffering. To live through this nightmare of fear that was with her at every instant, she had to toughen her will and spirit, to develop her outlook so that she had the stubborn, unyielding courage of a soldier. Finesse, subtlety, broadness of view were no good here. She had to learn how to ignore the probable, and deny what reason said was true and right. She had to believe the unbelievable. A closed mind, a rejection of all doubt and speculation—these were the virtues she cultivated. When they were brought over into the field of politics (as they were during the war) they became deadly vices.

---

8. Pierre Gilliard, *Thirteen Years at the Russian Court* (New York: George Doran Co., 1921), p. 51.

9. *Dnevnik...Nikolaia*, p. 270, September 8, 9, 10, 1904.

10. As honorary colonel of a Tatar regiment Alix had a right to use this title.

The pain of this tragedy caused the family to live with periods of gloom. No outsiders knew the cause, few observed the royal pair close up. One of the few who did was Pierre Gilliard, the tutor, who at times noted the strange behavior of the children when Alexis was reported to be ill.

> My pupils betrayed [their sorrow] in a mood of melancholy they tried in vain to conceal. When I asked them the cause, they answered evasively that [Alexis] was not well. I knew from other sources that he was a prey to disease which was only mentioned inferentially and the nature of which no one ever told me.[11]

When Gilliard first saw the child he took note of the strange actions of Alix.

> I saw the Czarina press the little boy to her with the convulsive movement of a mother who always seems in fear of her child's life. Yet with her the caress and the look which accompanied it revealed a secret apprehension so marked and poignant that I was struck at once.

In going through the memoir literature of persons who were close to the court, the reader is struck by the widespread ignorance about this illness. Members of families cursed by it knew little about it. In fact, Alexis's case seems to have given the disease more world-wide attention, after his death, than it ever had before. But while he lived, few knew the nature of his disease; it was not until 1912 that a number of persons became aware that he had a serious illness. No one knew that he had manifested the symptoms of his malady when he was only a few days old. It cannot be said with certainty whether Rasputin knew anything about hemophilia.[12]

Alix grasped clearly that the disease of the heir might cause important complications in the dynastic problem. If the knowledge got out that Alexis was incurably ill of an affliction that most likely would end in death then the royal pair would be where they were before his birth. She was convinced that many of the members of the imperial house were cherishing dreams of taking the throne, and awareness of the fact that the

---

11. Gilliard, p. 26
12. R.K. Massie, *Nicholas and Alexandra* (New York: Atheneum, 1967). This book contains authoritative material on the technical aspects of hemophilia and a searching discussion of the psychological impact of the disease on the royal couple.

boy might die would only arouse their unwholesome desires and would also tend to weaken the position of Nicholas. For these reasons it was decided that the illness would be treated as a state secret confided only a few doctors at court.[13] It was well kept; Pierre Gilliard, a man who visited the family for several hours a day five times a week, did not learn about it until 1912. When other members of the court and society wrote their memoirs after the revolution they made all kinds of guesses as to when they thought the illness first became apparent. Almost all guessed incorrectly.

In hiding information about the illness of Alexis the imperial couple hurt their cause and contributed to their isolation. What consolation and sympathy they might have received from some of the members of the family and from society was sacrificed for what they considered to be the more important question of the safety of the succession. The actions of Nicholas and Alexandra became a puzzle to all.

§

In the mind of Alix there was a close connection between the fate of Alexis and the fate of Russia. Besides suffering the anguish of a mother, she was troubled by the political issues. Anything that provided her with comfort and reassurance in these matters was much desired. In these things Gregory spoke the kind of hopeful words she wanted to hear. It has been believed that he convinced her that there was a connection between his presence near the throne and the safety of Alexis. Although Alix alluded to this idea she probably did not take it too seriously, but she obviously felt that the prayers of Gregory helped. Nicholas did not believe that the presence of the *starets* was any guarantee of the well-being of his son.[14] When the boy was going to the Stavka (the military headquarters) with his father in 1916 he suffered an attack of bleeding as a result of a minor accident on the train. Nicholas sent a series of cryptic telegrams to his wife describing the boy's condition. There was no mention of Rasputin or his prayers, although the tsarevich almost died.[15] The empress,

---

13. Gilliard, p. 31.

14. Nicholas II, *Secret Letters*, pp. 174-77, October 20, 1912. He did not mention Rasputin's alleged aid.

15. Nicholas II, *Letters of the Tsar to the Tsaritsa*, 1914-1917 (London: J. Lane, 1919), pp. 116-17, December 3, 4, 1916.

however, was certain that the prayers of Gregory had helped.

She had been convinced of his power of intercession by an incident that occured in 1912 at Spala, the imperial hunting lodge in Poland.[16] As a result of a blow on his leg Alexis was stricken with deadly bleeding. A crisis developed and his condition rapidly grew worse. All of the court doctors were summoned, but they confessed that they could do nothing. The tutor Gilliard was a witness of some of the hidden aspects of the terrible ordeal that followed. Still trying to keep the illness a secret, Alix and Nicholas attempted to carry on with the usual entertainments and receptions, not revealing the anguish they suffered. Gilliard reported how one night, when the screams of the boy were heard over the entire rear of the chalet, the mother came hurrying through a darkened hallway from the brilliantly lighted reception room. Gilliard drew back into the shadows and observed her "distracted and terror stricken look" as she rushed by unseeing. A short time later she returned to the reception room with the mask of a smile on her face. The doctors soon reported that the tsarevich would probably die and advised that carefully worded communiques should be issued preparing the public for the tragic news. The last rites were administered, and in all the churches of the empire prayers were offered for him. In despair Alix watched the life of her son ebb away. Then she rallied and refused to believe that the worst could happen, and she warned Nicholas not to issue the announcements. Nevertheless, a few vaguely worded bulletins requesting the prayers of all the faithful were issued, and the tsar steeled himself for the tragedy he thought was surely about to strike. Alix had more faith than her husband that divine intervention would save the boy. This was the final test of their faith. Who would be the stronger? In desperation she telegraphed Gregory in Pokrovskoe imploring his help. For him this was a moment of great opportunity and great danger. Paracelsus, the sixteenth century alchemist and physician once said that the healer's greatest power is in his own faith in his ability. Gregory had this. His career was staked on his answer to Alix. Playing the part of the holy man curing at a distance, he seemed to work a miracle. His soothing words were completely indifferent to the danger the boy faced, as if he were absolutely sure that all would be well in a short time. "The sickness does not appear to be dangerous," came the casual words. And next the insult to his competitors: "Don't let the doctors tire him." He added that he was praying and that all would soon be well. On

---

16. Vyrubova, pp. 90-93. Mosolov, pp. 150-51. Spiridovitch, *années*, II, 288-90. Buxhoeveden, pp. 132-33.

the next day Alexis suddenly improved, and shortly after he began the long period of convalescence. It was pointless for the doctors to indicate the importance of chance in the matter of coagulation. Once again Alix was able to point out to her husband that in perilous days his faith was weaker than hers. He was silent. Gregory now enjoyed her confidence.

VI

A *STARETS* ASCENDING

Gregory, a religious adventurer at court, at first had no clearly defined ambition or sense of calling; events swept him along as they swept along the other persons in the drama that featured him as the protagonist. Later, the empress supplied the ideology that was supposed to explain his presence and she also defined the position he was to fill. He understood approximately what was wanted inasmuch as some of her ideas corresponded roughly to his own simple outlook, and in his fashion he tried to play the assigned part, but in 1905 he thought of himself only as a peasant who had the good fortune to meet the tsar in his palace.

His rise to a position of importance at Tsarkoe Selo was slow; neither he nor the imperial pair realized what was happening. He was never able to manipulate or dominate events; he had to wait patiently to discover what factors operated in his favor and what were the situations he could elicit in order to ingratiate himself at court. A few of these factors (the hemophilia of the tsarevich, for instance) have been understood by observers, but some factors (including Gregory's alleged cunning in fostering his own career) have been much exaggerated. At an early date he perceived that it was important to be useful to the royal couple, especially to Alix, and he cultivated his cause whenever events offered him a chance.

He sensed that it was one thing to be received at Tsarskoe Selo and it was another thing to be welcome as a guest there on a permanent basis.

He established his importance through his personal relationship with the couple. He never held any post, nor was he ever given a sinecure. In many ways he was like the other visitors who came to court. Several of them could have boasted of intimate and frequent contact with the couple over short periods of time. But Gregory became unique because he was able to establish himself in the eyes of the empress as an independent man who represented no group or person but himself. By fortune, he also fit into her ideas of the Man of God, the Friend of God, who she hoped had come to help her save Russia.

Only Philippe briefly approached his degree of ascendancy at court. No one at the time, including Gregory, realized that the key to the position of paramount importance at the palace rested not with him who got close to Nicholas but with him who could win the trust of Nicholas' wife. The tsar was subject to moments of doubt and hesitation concerning his advisors, but when Alix found the Man of God most of her doubts dissolved. At first she preferred to observe Gregory at a distance because experience had taught her caution; many of her friends had proved to be unfaithful. In 1907 something happened that helped him to begin to acquire some of the all-important ingredient of independence. At this time Anna Taneeva, who had appeared at court earlier, became the friend of the empress.[1] Anna was the daughter of a distinguished family that had served the tsars for five generations. Her father, Alexander Taneev, was a noted musician and friend of Peter Tchaikovsky; he had turned courtier, held several important posts under Nicholas II, and was a member of the Council of Empire.

At first glance Anna seems to have defied all of the laws of genetics. She was a sport, totally different from the other members of her illustrious family. However, during her adolescence she suffered from many serious illnesses that may have affected her mind. Whatever the cause, she was unbelievably, freakishly naïve. In her memoir she could write with straight-faced candor: "The Emperor's combined billiard and sitting room was not very much used because the Emperor spent most of his leisure hours in his wife's boudoir." Anna was plump and moon-faced with a small mouth and huge blue eyes, always open, staring ahead with the quizzical, simple curiosity of a child. And like a well-behaved child she was shy and quiet, but when spoken to she replied in a stream of words

---

1. Buxhoeveden, p. 210. Dehn, pp. 47-50. Elizabeth Narishkin-Kurakin, *Under Three Tsars* (New York: E.P. Dutton Co., 1931), pp.186-212. Spiridovitch, *Raspoutine*, pp. 64-68.

delivered in a high-pitched voice, with a monotone that most listeners grew weary of very quickly. The ideas that tumbled out of her mouth were sometimes not connected in a rational order. They were strung together in the random fashion of a child who is interested in the amazing and curious rather than the important and meaningful. She had an excellent memory and retained facts well—without having a clear idea of their significance. Her sharp eyes and ears saw and heard all, and so she was a reliable reporter—provided one knew how to interrogate her and did not expect her to judge what her senses had taken in. The empress once dispatched her on a train trip with Rasputin and several court ladies to Pokrovskoe. She was supposed to observe and comment on his conduct and the attitude of his neighbors toward him. Rasputin, who knew how to be theatrical, let her observe him and some friends fishing with nets in the river. All at once the excited Anna, who for years had thought Gregory was a saint, realized what she was seeing: a recreation of a scene when Christ was with his disciples, the fishermen. On the train Rasputin became amorous with one of the ladies. Although she repulsed him he persisted, and later he tried to climb into her sleeping accommodations. She feared a public scandal so she silently fought him off. What he took for lady-like coyness whetted his appetite. When Anna looked at them he stopped, but then in a minute he would renew his attack with increased ardor. Anna could see only brotherly kisses and caresses. In raptures she wired Alix: "We live in blessedness."[2]

Anna was rejected by her family and their circles. The Empress Dowager and the Grand Duchess Elizabeth, wife of the Grand Duke Sergei and sister of Alix, more or less adopted the pathetic girl, and there brought her to court in the winter of 1901 when she was seventeen. It was there that Alix first saw her. She whirled through her first season with an endless round of balls, receptions, dances and teas, but at the end she was stricken with typhus. In a fever-induced delirium she saw a vision of Fr. John Kronstadt, who promised her she would live. Such simple-hearted faith made the smart set at court titter with amused scorn, but this was the very thing that attracted Alix to her. Once again the powerful maternal feelings of Alix were aroused, this time by the piteous condition of the girl who was almost an orphan, mocked and laughed at by the world because of her unsophisticated faith and lack of social grace. Anna quickly detached herself from the Empress Dowager and the Grand Duchess Elizabeth and became a close friend of Alix, who sponsored her rise through

---

2. Ibid., p. 95-103.

the lower ranks of the court ladies. In January 1903 she was given the diamond-studded *chiffre* of the maid of honor, which meant that once a suitable husband could be found for her she would have permanent entry to all court functions. Two years later she was appointed *Fraülein* in the personal suite of the empress; in the summer of 1907 she joined the imperial family on their annual cruise among the Finnish islands; and in the winter of 1908-09 she accompanied the family to Livadia. These last two honors meant that she was considered closer to the empress and emperor than the cousins of the tsar.

Anna's strongest characteristic was a selfless devotion, a full giving of herself and her allegiance to her protectress. She was completely dependent on the empress and could not be corrupted. However only a married woman could hold full court rank, and the strong-willed empress intended to leap over this barrier quickly. So eager was she to have Anna at her side that she hastily arranged a marriage for her with a young naval officer, Vyrubov, who had become mentally unbalanced as a result of his experiences in the Russo-Japanese War. So Anna was wed in April, 1907. But in a few months the match ended in separation, and at the end of the year it was officially dissolved. Vyrubov later recovered, married again, and had two children. After the revolution of 1917 Anna proudly boasted that doctors appointed by a investigating commission of the Provisional Government testified that she was correct in insisting that the marriage was never consummated—nor could Rasputin have been her lover. However, the misfortunes of 1907 touched off prolific springs of compassion in the empress, who was more determined than ever to help and protect the girl. The two became inseparable, and Anna was treated like an elder daughter in the imperial household.

No one at court had bothered to cultivate the friendship of Anna; she seemed to be an unlikely candidate for any kind of influential post. The favor seekers suddenly found to their surprise that a frumpy girl, a nobody, had run off with the most desirable position in the palace. Moreover, this nobody refused to play the game that was always played at court—she would not share with anyone what she heard and saw. Information or gossip was the commodity these people dealt in, it was the payment they got for help they had given to a courtier. But she had been discovered and raised up by the empress and therefore she was beholden to no one.

She was given use of the "Maisonette," a small and uncomfortable residence a few hundred yards from the imperial palace. There she lived and received Gregory whenever he visited Tsarskoe Selo. When the empress wanted to see him she went to the Maisonette; at the palace he would have had to meet the police and various secretaries and to place his name in the appointment ledgers. In Anna's house the empress could have the privacy and informality she wanted. It was there that the three members of the bizarre alliance, as poetess Zinaida Hippius called it, met to chat about religion and politics. It was to the Maisonette that Nicholas went for his rare meetings with Rasputin.[3]

Anna was indispensable as an intermediary between Gregory and Alix. Her fidelity was not a matter of being close-mouthed, for she was more than discreetly silent. She was willing in the name of loyalty to have almost no contact with society, and to accept the contumely this involved.[4]

Gregory derived substantial benefits from the arrival of Anna at court. However, their first encounter involved him in a dangerous dilemma. In 1907 he was asked to comment on her proposed marriage. At this time he was still friendly with the Montenegrins. They were alarmed by the fast-rising star of the parvenue maid-of-honor, and they were eager to put an end to her career or at least to slow its progress. Gregory's influence with the girl was likely reckoned to be great enough to cause her to delay the marriage if he advised against it. Such advice would have been an affront to the empress, who was the matchmaker. The same advice would have pleased the duchesses since an unmarried woman could not take full court rank. Rasputin chose to predict what his common sense told him—that the couple would not be happy together since Vyrubov's mind was still in shock. Militsa, who had introduced Anna and Gregory earlier in the year, was pleased—for the moment. In the eyes of Alix and Anna, following the tragic outcome of the match, the entire episode appeared as one more instance of his ability to see into the future, a gift that marked him as a Man of God.

From the outset Anna liked him and spoke well of him to the empress. Gregory on his part always showed consideration for the unfortunate girl. To Alix this was an additional sign of his goodness. From the time Anna went on the first cruise with the family on the *Standart*, the imperial yacht, she was reviled by society, which waged war on her reputation. The Montenegrins, having given up their favored place at court to the newcomer, resented her and joined the campaign against her. When it became obvious that Gregory was siding with Anna, they turned their

---

3. Zinaida Hippius, "La maisonette d'Ania," *Mercure de France* (August 1, 1923), p. 611.
4. Julia Cantacuzene, *Revolutionary Days* (Boston: Small, Maynard, & Co., 1919) p. 25.

wrath on him. Alix watched all these events. They could be explained by her notion that the just men of the world were hated by the wicked. As the Friends of God had said, the righteous could be known by the enemies they made.

At the tail end of this episode was a small but important happening: Gregory ended his friendship with the duchesses and was ever after able to boast that although he had friends he had no sponsors. He wore no man's livery. He was independent, the position he needed most if he were to become trustworthy in the eyes of the empress.

When Fr. John of Kronstadt died in December 1908 a potential danger was removed from Gregory's path. "The All-Russian pastor" was a friend of the prelates at the St. Alexander Nevsky Monastery, and later, when they tried to destroy Gregory they most likely could have relied on the help of John, whose words might have had some weight at court.

§

Gregory first achieved a public reputation in connection with his activity in the state church. Once the belief was widely accepted that he was a power in the church, the part of the public that was aware of his existence quickly presumed that his influence also extended to politics. It was a simple matter to conclude that he was a new Pobedonostsev who used a position in the church to dominate large areas of government. Gregory's lack of interest in the organization of the church and its hierarchies would have normally caused him to be little concerned with ecclesiastical affairs, yet events swept him along and by 1909 he was becoming involved in matters where church and state met. In this year his path once again crossed that of the Russian Savonarola, the monk Iliodor, who since the revolution had been leading an adventurous life in the public eye, seeking a ministry that suited his dreams and talents. Anyone close to the controversial monk was bound to share some of the limelight with him.

Iliodor at first preferred the life of an individual crusader, but for several years he had associated with the gangs of Black Hundreds and their leader, Dr. Dubrovin. When the government later began to withraw much of its support from this movement, Iliodor found himself tied to a declining power and as a result was forced to return to his role as lone publicist and agitator of reaction. He concentrated his fire on landlords, threatening them with violent expropriation if they did not give their lands freely to the peasants. In the pages of *Russian Banner*, an organ of the Black Hundreds, he hurled anathemas at the officials of the government and the Holy Synod for siding with oppressors of the poor. Izvolsky, the Over procurator of the Holy Synod, at the end of 1907 pressed Nicholas to unfrock the violent monk, but Nicholas read the articles and observed that the hierarchs Iliodor attacked by name were guilty of the offenses he charged them with; therefore, he was for the moment to be left alone. Iliodor could not believe that Nicholas on his own could display such understanding and generosity so he accepted Gregory's claim that he had intervened and influenced the tsar to be gentle. Throughout most of the Iliodor episode the tsar found himself in an embarrassing position—like the village policeman who tries to look the other way when he sees a small boy committing a nuisance. But the tsar could not stand idle too long while Iliodor lashed the bureaucracy with words that more and more appealed to violence. At the end of 1907 Iliodor was advised to go to the Pochaev Monastery in western Russia.[5]

He shared a belief common among lovers of autocracy that Nicholas II was a good man but was surrounded by a notorious "wall" of bureaucrats that prevented him from hearing the truth. During the next few years as his arguments with the authorities increased, he became obsessed with the notion that he could save himself (and Russia) if he could pierce the "wall" and get a personal interview with the tsar, who would then succumb to his spell-binding words. Dubrovin had bragged of his meetings with the emperor. While visiting the capital in May 1907, Iliodor asked the leader of the Black Hundreds to use his influence to obtain an audience for him. When Dubrovin failed the two men quarreled in public.[6] Iliodor denounced Dubrovin and went off to sulk and dream of finding a way to present his case at court. In the meantime, fiery writings and speeches poured from him.

He was next ordered to Tsaritsyn on the Volga. This town was a center of revolutionary ferment, and the authorities thought that the factious monk would probably destroy himself there. The peasants of the region were known to be still sullen and hostile after the events of the summer of 1906 when many had taken up arms against the military columns sent to pacify them. Iliodor preached his ideas and added to them a stream of curses against the local officials. He quickly won a mass following among the peasantry and the townsmen who helped him by contributing money and labor in building a four-story monastery-fortress in

---

5. Trufanoff, *Monk*, p. 50.
6. Bogdanovich, pp. 248-49, May 4, 1907.

which he could hide. As long as he continued to attack the local government, he was bound to experience the wrath of the officials. He continued to dream that he would be able to do anything if he could get to the tsar. In September, 1908, Dubrovin told the tsar that Iliodor was another John the Baptist, but Nicholas expressed no desire to see him.[7] Dubrovin could do no more for his friend. Depressed, Iliodor then turned to another friend, his superior, Bishop Hermogen of Saratov. Hermogen was a well-known orator, ascetic, and leader of important groups within the hierarchy. For several years he had been a leader of the Black Hundreds, and locally he had gained fame for his criticism of Tatishchev, the governor and friend of Stolypin. Although he was a highly placed and influential bishop, Hermogen had to admit that he could not secure the interview for his friend. As a gesture of desperation, Iliodor thought of invading the palace. Sometimes it seemed to him as if only God had the necessary power to get him access to the tsar.

The devils who were out to destroy him were about to catch him. In the spring of 1909 the authorities had succeeded in persuading the Holy Synod that the time had come for a final solution to the problem of silencing Iliodor. The monk feared that in the next censure he would be sent to a remote monastery where he would be a virtual prisoner. Without any definite plan in mind he went to St. Petersburg to look for help from Theofan, rector of the Academy, confessor of the empress, and sponsor of Gregory. A year earlier Theofan had gone to Pokrovskoe on an official mission for the empress. She wanted him to investigate Gregory's past. He reported that all the rumors she had heard about him were false. However, when Iliodor met him in 1909, Theofan was beginning to go over to Gregory's enemies. His influence at court was slipping and as a result he could do little for Iliodor. The unhappy monk was supposed to be in Minsk in White Russia where the authorities hoped he would preach against the large Jewish population of that town. It was Nicholas's idea to deflect him from his assaults on officialdom and turn him against Jews. Rasputin warned Iliodor that he had better carry out the wishes of the tsar and depart. At the critical moment in the Lavra at St. Petersburg in the spring of 1909, Gregory strolled into the room where Theofan and Iliodor were talking. After listening for a few minutes he casually offered his help. Shortly before this Iliodor had approached the Holy Synod directly asking that it rescind his posting to Minsk, for him a stepping-stone to oblivion. He was rejected. It was Easter time and all

---

7. Ibid., p.274, September 6, 1907.

churchmen were busy with duties; they had no moments to spare for a monk in trouble. Although there was little hope, there was nothing to lose in accepting Gregory's offer.[8]

§

Iliodor was trying to save a career; Rasputin was trying to build one. During the past few years he had been gradually ingratiating himself with the empress. Everyone was too distracted by the momentous events of the past few years to pay much attention to an obscure *starets* who was occasionally appearing at court. Even the personal security police had not spotted his first comings and goings, and when they suddenly realized he had seen the empress they made a hasty check on him. However, when the country was stabilized after the elections to the third Duma in February, 1907, attention once again began to turn back to the court. If the tsar was going to remain on his throne after all that had happened then it was important to know who was offering him advice.

Gregory was denounced for the first time at the end of 1908. Several members of the court had taken note of his presence, and had become increasingly apprehensive lest his name be linked with the emperor and empress. Before the revolution little harm was done by the presence of the religious prophet, the dwarf Mitia, who was sometimes a visitor at court.[9] Only a small number of people knew about him and discussed him. But after the revolution there was a large amount of freedom of the press, and therefore the newspapers could spread word of such things among tens of thousands of people in a few days. If there was anything suspicious or even ludicrous about Rasputin his presence could do untold damage to the monarchy. This was the beginning of the besmirch-the-monarchy charge, later used against Gregory many times in conservative circles.

To help him Gregory had an assortment of friends. But the first group was beginning to desert him. The churchmen at the Lavra and academy, the Montenegrins, and others were chagrined at his growing independence. In addition, he was no longer able to hide the debauches which shocked his clerical friends.

---

8. Trufanoff, *Monk*, p.96.
9. [Editor's note:] Trufanoff (a.k.a. Iliodor) describes Mitia as a "cripple" who "could hardly articulate," but would speak prophetically through interpreters. Trufanoff (ignor-

After he had served his apprenticeship on the high social and ecclesiastical levels he began to make contact with other groups—businessmen and certain elements of the petty bureaucracy, both classes trying to find ways to operate under the new dispensation of a constitutional monarchy where their needs were supposed to count for something. But as yet there were no private or public institutions to serve these needs. The new acquaintances of Gregory were the vanguard of a type which later swarmed around him. They were the disreputable entrepreneurs who specialized in the shoddy deal or the small speculation. They were the little merchants, ambitious but low-ranking officials, actors, part time journalists, feuilleton writers and scribes, persons of no definite occupation, who loitered around the edge of the reputable world of government and society, floaters, adventurers who lived on the edge of the law and sometimes slipped over that edge. Although they did not have the power to place Gregory at court they might gain advantage from his being there. They watched from points close to him, fascinated, their predatory instincts telling them they had a stake in his career. Quick to recognize a rising star, they cultivated Gregory's acquaintance and were glad to assist and protect him when he disported himself in their midst or indulged in riotous behavior in the restaurants of St. Petersburg.

At court Gregory had few friends, most people hated him. He had sympathizers among a few minor officials such as the courtiers Loman and Sablin. However, his foes were formidable: Prince V.N. Orlov, assistant chief of the emperor's military chancellery, a powerful leader of military and aristocratic elements in the capital; his assistant Dreteln, popular at court because he was a charming raconteur; V.A. Dediulin, Commandant of the Palace and aide of the emperor; Madame Naryshkina, eldest Mistress of the Robes and an important leader of the nobility.[10] The police were also in the camp of Gregory's opponents. The Okhrana expressed alarm over the presence of the *starets* because they found that although he was politically safe, he moved in strange circles and met many people with all sorts of suspicious backgrounds. They generally gathered in a number of salons interested in pietism and influence-peddling and were therefore not easy to watch. A.V. Gerasimov, speaking for the police, and Dediulin, speaking for the court, asked Stolypin to support an assault on Rasputin. The premier agreed and charged Gerasimov to assemble the

---

ing Pobedonostsev) suggested Mitia's counsel contributed to choices made leading up to the 1905 revolution. He was fourth in the line of eccentric spiritual advisors to the tsar. *Monk*, p.170-178.

10. Narishkin-Kurakin, p. 195.

dossier for presentation to the tsar. The report included material from the Tobolsk consistory hearings and recent surveillance of Rasputin carried out in the capital. Stolypin personally brought the material to Nicholas and asked him to order the *starets* to Siberia.[11]

The attackers considerably underestimated the degree of influence that Rasputin had with the empress. But they also committed other blunders. They did not know that he had long ago successfully neutralized the Tobolsk papers by freely admitting the truth of whatever was compromising in them and confessing that he had once been a great sinner. His sinful past and virtuous present were the most potent factors that recommended him to Alix. Her own husband had come by a similar path—with much less sin, of course—to the present. She probably suspected she provided some of the grace which saved both men. But how could the attackers know such things? The inclusion of the materials of police surveillance accomplished nothing. Many times in the past Nicholas found that the police were capable of falsifying evidence even when they new the tsar would see it. The imperial family did not like the police and had little respect for them. The fact that the initiative for the attack came from persons at court also served to discredit the assault in the eyes of the tsar. Alix quickly came to the assistance of her favorite by dispatching Theofan, the unworldly monk, to Siberia to live for two weeks with Gregory in Pokrovskoe. Theofan reported back in words glowing with praise for the *starets*. Gregory was saved temporarily, but he knew that powerful enemies were watching him, their forces still intact. They might strike again.

In the spring of 1909 that moment came. His enemies discovered that Theofan could be won to their side. The naive cleric had preferred to think well of every man. As long as he thought Gregory's sins were buried in the past he supported him. But now disillusionment had set in. He was very much impressed by a barrage of facts describing Gregory's nightly rambles in the city. He was aggrieved at his own gullibility, but hoped that Gregory still might be saved. Although he saw that Gregory was not the holy man he pretended to be, Theofan thought he was not guilty of the widely accepted charge that he was a *khlyst*. Therefore, the task of redemption might not be as difficult as others feared. Theofan did not break with his protégé at once; instead, he warned him to stop his carousals. As a sign of his displeasure and as a warning, Theofan refused at Easter to give Gregory the traditional forgiveness for wrongs committed during the year.[12] No one knew better than Gregory that it was impossible for

---
11. Ibid., *Raspoutine*, p.96.
12. Spiridovitch, *années*, I, 296-98.

him to mend his ways and to conform to the demands of Theofan.

He sensed that he was about to lose a valuable supporter whose desertion might even shake the faith of Alix. Up to this time he had been accused of being a debauchee, but when the learned rector had deserted him, he was left open to the *khlyst* charge. He was not certain that he could win in a test of strength between himself and the monk. Theofan was the keystone in the alliances Gregory had built up in several church circles, and if Theofan turned against him Gregory realized that many other friends would disappear too. They were the very people who normally could not be won over by Stolypin because they did not like his politics, but they would listen to and respect the word of Theofan. Neither Gregory nor the empress placed very high value on the support of the officials in the church, but if his friends there deserted him, then the courtiers, the police, Stolypin, and all sorts of other persons might be emboldened to make common cause in a grand assault on him. Who was certain that Nicholas could resist the demands of such a coalition if it asked that Gregory be exiled?

Iliodor, who had been under a cloud of disapproval, was by no means an ideal ally, but Gregory was not in a position to wait for a perfect choice. Besides, in certain circles of the government there was a growing tendency to re-evaluate the meaning of Iliodor's activity. Disenchantment with Stolypin was increasing. Nicholas had begun to think that perhaps he had conceded too much in 1905, and the time had come for him to try to regain some lost ground. In politics there was restiveness on the extreme Right signalled by a growing desire to return to some of the ways of the pre-revolutionary era. Stolypin was a constitutionalist who warned the emperor that such an attempt to go back would eventually bring on a revolution. He made himself the bulwark of resistance to attempts to return to the old days. It was almost impossible to hurt him by direct assault. It was for this reason that his enemies approved the attacks by Hermogen on Tatishchev, governor of Saratov, and an appointee of Stolypin. But there were other ways in which Hermogen and Iliodor might be useful to the far Right. The monk was unusual because he was a popular leader who was at the same time anti-revolutionary. Essentially, he was in favor of Orthodox, autocratic Russia, but he sometimes hurled his poisoned barbs at friends of the tsar. At court there was speculation that he might be convinced or tricked to support the government. The depth of his convictions and the nihilistic lengths that he would go to in order to protect them were underestimated. Nevertheless, it was felt that he could be wooed, and if he were won over he would be valuable to the state and to those who wanted to see Stolypin injured in some way. Rasputin was probably aware of these currents of opinion and the reassessing of Iliodor.

An attempt was made to secure Iliodor's allegiance. Dediulin and his assistant, Spiridovich, were the first to receive him and to try to change him from a scourge into a supporter of the government. They lectured him on the need for discrimination and temperance in his denunciations. Time and its troubles had not mellowed Iliodor; he still sought the chance to talk to Nicholas. When he discovered the meeting with the two oficials was not a prelude to a confrontation with the tsar he stormed out.[13]

§

At this moment Gregory offered help. In the depths of his unhappiness Iliodor accepted. He waited resignedly but then grew restless. He heard that Gregory was visiting the Archbishop of Finland so he called the episcopal palace to find out what Gregory was doing for him. An official answered and said that Gregory indeed present but was napping and could not be disturbed. Iliodor was impressed. What happened next left him speechless. Madame Lochtina, a typical female hanger-on of Gregory, called on Iliodor the next day carrying an envelope of grey crackled stationary with the embossed emblem of the imperial eagle. It was an invitation to go on the very next day to Vyrubova's Maisonette, where Gregory, Anna, and the empress would receive him. The audience took place on April 3, 1909 at 3 P.M. "for fifty minutes" as Iliodor later noted with great meticulousness. Gregory's influence was awesome.[14]

His skill was also impressive. For him the entire episode offered many advantages. It demonstrated to the courtiers the extent of his power; they would have to think twice before attacking him again. Although Iliodor was a minor thorn in the flesh of the government, his power as an irritant might increase. Nicholas hated to deal with him. Gregory reassured Alix that Iliodor could be managed if the right approach were used. The matter was trifling enough for Nicholas to permit her to handle it, yet if she were successful it would be a good entry into politics. And

---

13. Ibid., *années*, II, 44-45.
14. Trufanoff, *Monk*, p. 101.

so the incredulous monk was brought to the palace where he met the threesome. He was told the tsar would repeal his exile but that certain conditions had to be met: he had to soften his attacks and sign a statement the empress gave him in which he would promise to behave himself in the future. If he agreed to these conditions the repeal would stand. In addition, hints were dropped that there would be other rewards for him: a bishop's hat, perhaps, and his coveted interview with the tsar. These last two touches had the marks of Gregory's caution and guile. He knew how dear to the monk's heart was this meeting. Iliodor had a crazy dream of winning the absolute confidence of the tsar, who then would make him a kind of official government prophet and tribune of the people, with vast powers to censor the rampaging officialdom and to stop them from tormenting the poor. Gregory did not have the power to bring him to the tsar, but Iliodor did not know this. Instead, Gregory placed all the pressure of responsibility on Iliodor by tying the chance of an interview to the monk's good behavior. He knew Iliodor well; there was not much chance he could maintain the terms of the agreement for long.

The talk with the emperor was not all the monk had hoped for, but it was better than talking to more bemedalled police officials whom he hated. It was a temporary truce. After this his sermons contained less vituperation against officials. In return for his moderation he found that the spectre of immediate censure was lifted. He made a partial comeback from the verge of destruction—thanks to Gregory.

When he thought about what had happened to him on that memorable day at the Maisonette, he was impressed with the fact that although he had a number of important friends it was only Gregory who had brought him to the empress. He was struck by the offhand way that the meeting had been prepared once Gregory made his offer. He did not know, of course, that the empress was as eager for the interview as Gregory. Iliodor correctly interpreted the promise of a visit to the emperor as a threat as well as an offer, and in his memoirs he admitted that for the next two years he was intimidated by the apparent power of Gregory and his ability to carry out what at first seemed merely farfetched promises. It was hard to know what to believe about this extraordinary man. Was he a *starets*—or a devil?

Late in 1909 the police under Gen. P.G. Kurlov, an experienced administrator and the new head of the Okhrana in St. Petersburg, were renewing their investigations of Gregory's activities, and it was rumored that Stolypin was again prepared to show their reports to the tsar and to ask for the banishment of the *starets*. When he saw trouble on the horizon, Gregory thought it might be well to use his favorite defensive tactic: he skipped town, going to the only place he had friends—Tsaritsyn and Saratov. There he could dramatize his new connections in church circles, show how he had quieted the troublesome Iliodor, and display himself before the empress as a beloved *starets* hailed by mobs of people. In Saratov he could meet Bishop Hermogen for the first time. But Hermogen never warmed to him; however, Gregory had assisted Iliodor and for this reason the bishop accepted him as a useful ally.

Toward the end of 1909 Iliodor was growing restive with his restrictions. Stolypin was warning the tsar that the monk had to be shut up completely; he was preparing another *Pugachevshchina*.[15] The monk wanted to hit back. Gregory began to pursue him in order to make sure that the original impressions derived from the confrontation with Alix were properly amplified. Although Gregory knew Iliodor was a fool and would sooner or later ruin himself, it was against his interest for the end to come so soon. He needed Iliodor yet, and for a while longer he wanted to help Alix bask in the glory of her achievement.

Therefore, in November, 1909, he invited Hermogen and Iliodor to join him in Pokrovskoe, the place where he could usually present himself in the best light. For the next six weeks he moved about restlessly, usually in the company of Iliodor, between Pokrovskoe, Tiumen, Saratov, and Tsaritsyn. The high point of the drama came in his native village where he casually pointed to the signs of his success. He exhibited several shirts he claimed the empress had made for him. In his house he pointed to the more expensive furnishings and said that Alix, Anastasia, and Militsa had chosen them for him. He added that Anastasia was grateful for his help securing a divorce through the action of the Holy Synod. In making this remark Gregory was committing one of his occasional acts of daring. The divorce had been granted in 1906 when he was still unknown and powerless; in addition, Hermogen was close to several members of the Holy Synod and might have checked his statement. But before his friends had time to think he whirled them along with new presentations. When the mood struck him he dashed off telegrams to "Papa and Mama" at Tsarskoe Selo and received the usual prompt, friendly—and even obsequious—answer. This was one of his favorite stunts. The telegrams went to Vyrobova who replied using Alix's name. The ruse usually impressed observers. Gregory had more tricks: he said the tsar often begged him for advice—which he freely gave, although, he complained, the tsar did not always follow his wise words. He boasted he had raised up Sergei to his post as Bishop of Finland, he had made Theofan rector of the academy, and hinted that there were many more achievements but he did not want to be tedious. He suspected that somewhere in his soul, Iliodor yearned for a place in the sun. Gregory could not believe that any churchman could not be tempted with a promotion. That was what all of them seemed to live for, in his estimation. So he renewed the offer of a promotion to *archimandrite* (abbot) and added that later perhaps even a bishopric might be forthcoming. As a crowning touch he nonchalantly handed Iliodor a packet of letters and told him to keep them as souvenirs. These had been written to him by Alix and her daughters in the prescribed submissive and affectionate style one was supposed to use in addressing a *starets*. Iliodor knew at once what a great favor Gregory had done him in giving him the packet because he could display them and prove that he had a friend who received letters on a household stationary of the tsar. The show was not over. A few months later in the capital Gregory introduced his friend to a high ranking courtier who addressed him deferentially as "Father Gregory."[16] Then he let him see Vyrubova kneel, in one of her almost hysterical outbursts of religious emotion, and kiss his hand.

Such sights overwhelmed Iliodor, and convinced him that Gregory was a person wielding great power. At the same time he began to hear disturbing stories of Gregory's personal life. The *starets* himself was swept up in the excitement of his tour among the people in the Tsaritsyn-Saratov area and he began to kiss and embrace women when they came to him. In private he went further. A nun, Xenia, and the wife of a shopkeeper both complained to Iliodor that the holy man had tried to molest them. In Pokrovskoe Fr. Peter and his assistant told Iliodor about the past of the *starets*. Iliodor was disturbed, but for the moment he decided to say nothing—probably because he needed Rasputin. Besides, there was much excitement, turmoil, and movement around Rasputin; it was hard to stop to think about what was happening. Even if some of these stories were true, the power of the man was a thing to be reckoned with and, whenever possible, to be used, Iliodor thought. Unlike many persons who got close to Gregory, the monk at this stage did not underestimate him. He was inclined to take at face value all the trappings of influence, and in his speeches and writings he spread the word of Gregory's important connections and about his ability to help those who befriended him. From this moment the legend of Rasputin's influence took wing. The small amount of truth was embellished as it travelled. Those who sym-

---

15. A massive rebellion in the 1770's.
16. Spiridovitch, *Raspoutine*, p. 127.

pathized with Iliodor's story believed it, and with the passing of years his enemies believed it too because this legend seemed to be consistent with the few and imperfectly understood facts at their disposal.

The trip ended for Gregory with a series of victories. Most of the tales about him that were whispered in the circles of court, police, and church had not reached this region. The story of the interview was, however, well known. Iliodor introduced him to all prelates as a close friend of the imperial family. Gregory then temporarily stepped into the new and unfamiliar role of popular idol of the crowds. Thousands of peasants trailed him in processions through streets of the cities, everywhere people pressed around him imploring his blessings, or hoping to catch a glimpse of him. Others stood for hours in the cold and snow outside railway stations or churches, waiting to petition him. Several church publications claimed that he was rumored to have performed miraculous acts of healing. The exultant Vyrubova brought these tales to the attention of Alix, and whatever the journals missed Gregory carefully supplied in a stream of telegrams to Tsarskoe Selo. When Iliodor returned to the capital he spread the word of the wondrous things he had seen with his own eyes during the momentous days at the end of the year.[17]

Alix took note of these events. At the end of December, 1909 Gregory gave small gifts to some people in a crowd of 15,000 in Tsaitsyn, and a few days later 2,000 people saw him off at the station for his return to the capital. Alix's respect for him was increased.[18] This trust and support was the only thing that saved him during the tumultuous days of the next two years when he was attacked from all sides.

His growing reputation tempted his foes early in 1910 to make another effort to drive him into exile. The organ of the Constitutional Democratic Party, *Speech*, indulged in a flurry of speculation, accusing him of having a bad reputation and of influencing affairs in the church. This was probably a variation of the theme that he had tarnished the reputation of the autocracy. Consequently, the press articles stirred up little interest among the opposition in the Duma, for in their eyes the monarchy was already tarnished, but for other reasons. However, several groups in the church thought the charge raised a serious issue. They believed he was evil and he would use his influence for a wrong cause. Now the old charge that he was a *khlyst* was revived. This would give all shades of Orthodox a reason to fear him and to resent his presence at court. The pro-*sobor* people believed that his presence was one of several reasons why

---

17. Trufanoff, *Monk*, p. 107.
18. Ibid., p.122.

the Holy Synod would not call a council or even discuss one. Rasputin had been heard to use the slogan, "There is an anointed tsar." During the first World War he used this cry to reassure the empress, and she in turn was supposed to use it to reassure the emperor that autocracy could solve all problems and needed no reforms nor help from any other institutions. What he meant by it in 1910 is hard to say, but his enemies concluded it was his way of supporting Caesaropapism.

Observers thought they detected at least one example of his helping clerics who shared his views. In 1908 Sergei, head of the Lavra, was promoted to Archbishop of Finland. Theofan in turn was appointed to the vacant post in the monastery. Iliodor had publicized the boast of Rasputin about his alleged role in these changes. Sergei was known to be unsympathetic to the idea of a *sobor*, and in his new post he was in a position to influence any inclination the clergy or Holy Synod might have had to request such a council or even to consider it as a serious possibility. As head of the academy, Sergei had not resisted the rise of Rasputin. He was one of the few who were not for long deceived by Rasputin but who thought he could be reformed. During the last two years of Rasputin's life, when he really had power to meddle in certain church affairs, Sergei was not afraid to denounce him. But in 1910 the linking of the names of Sergei and Rasputin tended to hurt the reputation of the churchman. Therefore his friendship lessened his influence when he called on prelates to resist a *sobor*. To neutralize Sergei and Theofan many churchmen were willing to accuse them of being Rasputin appointees. Sergei could not openly deny these rumors without causing displeasure at Tsarskoe Selo, but his silence tended to confirm the rumors. As a by-product of events he was little concerned with, Rasputin's reputation and importance were considerably enhanced.

Theofan must have suspected that the talk about Rasputin's power was untrue, but he had already concluded that the so-called *starets* was evil and could not be reformed; therefore, he had to be driven out by organized pressure exerted by all persons who realized what an incubus he was. The campaign drew support from church and court circles where there was less concern with the *sobor* and more concern with the way in which Rasputin's notorious reputation was hurting both monarchy and Orthodoxy in general. Theofan spent some time in efforts to convert Rasputin, trying to lure him into repentance. When he saw the effort was hopeless, he quietly went about gathering support for his campaign to drive him out. When he was seeking to change Rasputin's behavior, and was enlisting aid for his attack, the rector made no public statement of his revised opinions. As a result, when he publicly announced his opposition in the spring of 1910, his remarks were unexpected and puzzling—particularly to the imperial couple. To them the suddenness of the change smacked of opportunism. Shortly before he had made his open condemnation, Theofan's activities were reported to the empress by the Okhrana.[19] This contributed to her feeling that there was more guile than honesty in his actions. His reputation and position at court demanded that he cast himself in the role of the honest man, outraged and righteous, who denounced evil the moment he found it. Instead he chose to make his assault part of a sweeping campaign that included a report to the tsar by Stolypin and almost simultaneous denunciations by Sofia Tiutcheva, the tutor of the tsar's daughters, and the nurse Vishnaikova.

Iliodor nearly joined the plot now. He was in the capital in April, 1910, and he stayed at the house of the journalist G.P. Sazonov. He then talked to the disillusioned Fr. Benjamin at the Lavra. From there he went to Dean Solodovnikov of Vladimir. These clerics were part of Theofan's group united to attack Gregory. Next, Iliodor for the first time met some of Rasputin's clique and had a chance to see them in action. Sazonov, two English businessmen, a Professor Migulin, and several members of the Taneev family had concocted a crazy scheme to irrigate all of Central Asia. They had no money but they had Rasputin, who promised to take the idea to court.

Iliodor had once, probably about 1904 or 1905, planned with Hermogen to have Gregory ordained. As a priest he would have been under ecclesiastical discipline, which would have meant Hermogen. But Iliodor reported the *starets* was too ignorant to learn simple prayers. Gregory had easily learned much more than his prayers, but he had no intention of submitting to control. By 1910 Iliodor had abandoned the idea of ever influencing the *starets*. Now he wondered if he should help in destroying him.

Gregory was trying to keep Iliodor from joining the plot. Theofan conceived a puerile attempt to win over the monk. He had Fr. Benjamin, who was a leading polemicist at the Lavra, write letters to Iliodor, trying to induce him to desert his friend. The appeals sought to copy the style of Iliodor who was noted for his blood-curdling epithets. Iliodor was not yet ready to change sides. He gave the letters to Vyrubova who took them to the empress. Alix regarded them as evidence of the irresponsible wildness of Theofan and the Lavra monks.

---

19. Maurice Laporte, *Histoire de l'Okhrana, la police secrete des tsars, 1880-1917* (Paris, 1935), p. 234.

The ever-alert empress suspected that these incidents were not honest attempts of independent servants of the crown to deliver a warning. To her the attacks on her favorite assumed the proportions of an unwholesome and widespread conspiracy that included a number of persons for whom she had a special mistrust. The Montenegrins injured their reputations by associating with the plot. The lady of the imperial household, Tiutcheva, had for some time been having trouble in the palace. A member of a well-known artistic family, she was close to the Moscow Slavophiles (called derisively "Muscovites" in Alix's letters)[20] and was a super-patriot who had carped about the use of English and French in the royal household, where she wanted nothing but Russian to be used. She was also an amateur publicist for the *sobor*. Her uncle, the prominent bishop of Vladimir, spent ten years as Russian representative a the Vatican where he saw the benefits of an independent church. At first her demands were looked on as amusing and harmless nonsense. But she had overstayed her welcome and was about to be dismissed when she protested Rasputin's presence, complaining that he wandered in and out of the bedrooms of the tsar's daughters. Later, when she was fired the cry went up that Rasputin had done it.[21] The leader of the Octobrist Party in the Duma, Guchkov, was close to the Moscow Slavophiles. On March 8, 1910 after a political tempest in the Duma caused by a coarse speech of the reactionary delegate V.M. Purishkevich, the president of the Duma handed the gavel to Guchkov, who became the new president. The suspicious empress no doubt saw this as part of a plot, and she believed that Guchkov was behind the warning delivered by Stolypin and the actions of Theofan and the ladies of the imperial household. She suspected that these hostile forces were at work to undermine the monarchy; the incident involving Gregory was merely an excuse to cause trouble so their foul work could be done.

Throughout the summer the rector carried on his attack in the newspapers. Finally, in November he was ordered to give up his post and take the position of Bishop of Simferopol in the Crimea. This was a kind of gentle chastisement by the emperor. During the remaining weeks of the year the press devoted several articles to Rasputin; their reflective mood probably indicated the kind of thought that all of Rasputin's enemies were indulging in at the moment. For once there was a mood of

---

20. *PTR*, Guchkov, IV, 250-52.
21. Gregorii Shavelskii, *Vospominaniia poslednego protopresvitera russkoi armii i flota* (New York, 1954), I, 62.

sober reappraisal and judicious consideration of the facts. It was apparent that Rasputin had weathered another storm. There was speculation that his opponents may have underestimated him, had misconceived their mission in some way, and had overstated their case. Unfortunately, this fleeting moment of candor and realism on the part of the opposition was not brought to the attention of the royal pair because they spent the last three months of the year abroad, visiting relatives in Germany. There remained the need to find some kind of suitable explanation that would reveal how Rasputin survived so many attacks from so many quarters. It was finally agreed that he had great power over the tsar, that he could influence him in any way he chose, and it was further believed that Rasputin had the ability to make all sorts of appointments in the church. Not all the participants of the struggle of the preceding few years were ready to capitulate. From this time forward Rasputin had to live under recurring threats, and in almost every year there were a number of ritual warnings delivered to the tsar about him.

Whenever the warnings seemed on the verge of success, or whenever his name was mentioned too often in the press or court, Gregory left the capital and fled to Pokrovskoe. In 1915 and 1916, for instance, he spent as much time in this place or in travelling to and from it, as he did in St. Petersburg. Since each trip was accompanied by rumors that he was in voluntary exile or that he had left under orders from the emperor, his disappearance usually caused some of his enemies to lose interest in him. On several occasions his departure probably saved him from embarrassment, including the possibility that he might suffer genuine exile. He relied on Vyrubova to tell when imperial displeasure had reached the danger point, but she was not always capable of making this judgment. On several occasions he suffered humiliation and received suggestions that he depart.

In the spring of 1911 he suffered a disgrace of this kind when the irate Nicholas either ordered him out or declared such an order to be imminent.[22] Rasputin was caught unawares and he fled precipitously.

It all began with Iliodor, who was once again on the edge of rebellion. To keep him in line, Gregory continually dangled before him the promise of an interview and a promotion; as a result, even when he began to speak out again in the first weeks of 1911, the monk was not quite the firebrand he was of old. Stolypin, speaking through Lukianov, the Over Procurator of the Holy Synod, ordered Iliodor to be silent, but instead he

---

22. "Boris Nikolskii i Grigor Rasputin," *KA*, LXVII (1923), 157.

became more defiant. Stolypin again warned Nicholas that the monk was preparing a revolt. The tsar thought about the threat and decided that the monk should be removed to a monastery in Tula where he should be given a post equal in importance to the one he had been deprived of. When Iliodor resisted, Nicholas was puzzled about the next move that ought to be made. He was assured that any strong action by the government would most likely lead to violence, as Iliodor enjoyed the solid support of his people. Nicholas's views had changed somewhat since the last crisis caused by the fiery monk. The tsar was disenchanted with Stolypin. The empress already hated him and had been hounding her husband for some time to get rid of him. Iliodor's attacks were causing pain to Lukianov, Stolypin's friend, who was trying to gradually loosen the hold of the government on the church. It was the opinion of the tsar that much of the recent trouble and dissention in the church was the result of this lack of a firm hand at the helm. Other advisers reminded Nicholas of the upcoming elections. Conservative forces were preparing to organize the electoral campaign in such a way that a trustworthy majority in favor of the government could be assured. It was felt that the services of the Black Hundreds might be useful in rallying supporters of the government, and Iliodor was one of the genuine heroes and the most effective organizers of these groups. In addition, he was one of the darlings of the subsidized reactionary press that was expected to play a major role in the campaign. All of these considerations entered into the tsar's judgment about Iliodor's case, but he was still alarmed by the monk's noisy outbursts.

    Nicholas suspected that he might have been misinformed, that he did not have all the facts. Several of the persons reporting to him were known to be hostile to the monk. Nicholas called in a trusted aide, a fellow officer from his own Preobrazhensky Regiment of the guards, and sent him on a mission to Tsaritsyn to ascertain the facts. He told the officer, Captain Alexis Mandryka, that he appreciated the good work Iliodor had done and revealed that since the end of 1910 he had been receiving many telegrams praising Iliodor. On the other hand he was being assured by some of his advisers that the monk had no real following. Mandryka was empowered to make an estimate of the situation; if Iliodor was truly guilty of insubordination he was to be sternly warned and lectured by the officer. Mandryka carried out his mission in February 1911. In the company of the vice-director of police and the governor of Saratov "in full uniform and covered with medals and ribbons" he visited the monk and told him it was the desire of the tsar that he take the post of superior of the Novosil Monastery in Tula Province. With the mission completed the officer returned to Tsarskoe Selo and delivered his report in person to Nicholas and Alix. They listened sympathetically to the part of the report that concerned the monk. The emperor was convinced that he had a considerable popular following and could not be dislodged without a scandal, so he decided to try to be careful in handling the case.

But Iliodor was not the only one who suffered from this investigation. Quite unexpectedly Mandryka came upon evidence clearly showing that Gregory was meddling in the mission. Proof of this almost ruined Gregory. Vyrubova had warned him concerning the journey of Mandryka. The empress, who in a way was working behind her husband's back, probably instructed her to do this. Mandryka was a relative of a nun in charge of the Balachov Monastery near Tsaritsyn. Rasputin, who knew this woman, informed her of the coming of the captain and requested that she do everything possible to influence him in favor of Iliodor, and he added that he would appreciate her mentioning a good word about his own name if it were brought up. Mandryka learned of this tampering and had the luck to secure the first of Rasputin's telegrams: "one of your relatives has been sent on a mission to Tsaritsyn concerning our business. Try to influence him." Here was evidence that he had been trying to interfere with a mission ordered personally by the tsar. As a result of this, Gregory had to deal directly with the emperor's wrath. For the first time the empress was not in a position to assist him although she personally thought his intercession was merely impolitic. The emperor added this incident to a number of other irritants Rasputin had caused him lately.

Within a few days of Mandryka's return, Gregory made a last minute appeal to Stolypin for help. At the interview he gazed fixedly into the premier's eyes for several moments, but Stolypin later said he felt only an "indescribable loathing for this vermin sitting opposite me." His last hope gone, Rasputin fled the capital and began a long journey that kept him away for more than three and a half months on an extended pilgrimage to the Holy Land. In fact, he was away from St. Petersburg for most of the remainder of the year. His fall from grace was noted with much gratification in the capital, and most of society believed this to be the end of him. At first no one knew the cause of his trouble. The story of Mandryka did not become generally known until much later. The officer disappeared from court shortly after the mission, and it was assumed he had been banished for doing his duty. In the summer of 1914, however, he appeared once again; this time he was governor of Nizhnil Novgorod, a prestigious sinecure. But in 1911 gossip continued to spread the word that he too had been ruined for telling the truth about the powerful *starets*. One more imaginary victim was added to the list of those crushed by the all-powerful Rasputin.[23] The public missed the significance of the case. When the emperor found the first clear evidence that Rasputin was meddling in things that did not concern him and when there were no other issues at stake, then Gregory suffered a quick fall. Mandryka was not a police official, the Rasputin telegrams were not falsified or padded, the Flagellant charge was not brought up. Mandryka had stuck to his job and therefore had succeeded. Unfortunately, no one noticed that these were some of the ingredients needed for a successful attack on Rasputin.

Within a short time the attention of the nation was focused on more important things. Stolypin's hour had struck, and the country watched him fight his last legislative battle in which he sought to introduce local government in the western regions of the empire, but found heavy resistance in the Duman and got an unfavorable vote in the upper house, the Council of Empire. Then he suspended the Duma for a few days while he passed the bill by administrative fiat. The emperor was furious because he was forced to back his premier; Guchkov resigned his presidency of the Duma as a form of protest. In all this excitement Rasputin passed from the scene like a bad dream suddenly ending. But at the palace he was still remembered by Anna and the empress, who reverently read the letters he sent back while he wandered to Jerusalem. He still had friends in St. Petersburg and Siberia who remembered him as well.

Iliodor was now on his own. In a few weeks he was delivering intemperate speeches, and the Holy Synod once again ordered him out of his post. By this time he had so often escaped censure that he had acquired the reputation of a man who was almost immune from it. But now he had no friends or help. Even Hermogen could not assist him. In a short time he was transformed into a hedge-priest, running from the police whenever they closed in. He locked himself on a train and refused to leave. All passengers were removed, and for two days an engine with its puzzled driver ran back and forth on the same track while Iliodor fasted in a hunger strike. Several times in his flight he was caught by the police but escaped. Wearing a disguise he slipped into Moscow. Telegraph wires hummed as governors consulted Stolypin on how to manage the wild monk. The government was on the defensive; it looked ridiculous as it

---

23. This account of the Mandryka investigation is based on the references to it in Spiridovich, Rodzianko, and Gurko. Both Rodzianko and Gurko, who was an excellent historian, were able to consult the MS. Rodzianko's date of Rasputin's departure is a month too late. The former president of the Duma wrote in exile and in illness and is sometimes inaccurate in his dates.

was outwitted by Iliodor who ran about the country a fugitive from justice. The premier had to ask the tsar to defer the firing of Lukianov lest people think Iliodor had done it. Nicholas was thoroughly dissatisfied with the role of the Holy Synod in all this. He demanded a solution although he was not sure what could be done. Iliodor then fled to Tsaritsyn again.

Here things took a comic turn. He and his followers barricaded themselves for twenty-seven days inside the fortress-monastery. His demonstration of the impregnability of his position attracted the amused attention of the press and so-called "war correspondents" were sent to report on the "siege."

From afar Gregory tried to keep a hand in affairs so that he could maintain the fiction of his power. Several telegrams were sent to court advising leniency for his friend. Nicholas had wavered and then concluded it was best to leave Iliodor alone,[24] but Gregory was striving to create the impression that his opinion was being sought before governmental policy was made. Nicholas was finding out that without Gregory, Iliodor was much harder to manage.[25] When he finally retreated, Gregory was told of the impending order. Anna kept him informed. His telegrams from Jerusalem were thus dated before the order went out. In the summer of 1911 Gregory claimed that he had changed the tsar's mind.

Nevertheless, Iliodor was beginning to doubt that the *starets* had much power, and friends assured him that Rasputin had nothing to do with the capitulation. Shortly after Gregory returned to Russia, Iliodor was called to Tsarskoe Selo for the long-sought interview. After a talk he was invited to celebrate the liturgy in the presence of the tsar and court. The news of his proposed appointment as archimandrite was published in the official journal *Government Messenger*. The tsar was in an all-out effort to win over the monk. Iliodor was convinced that he would have little further need for Rasputin, who now seemed to be similar to a number of other cast-offs, former holy men who had flourished temporarily at court. Once again Gregory's incredible luck was with him. The monk was about to go down to ruin, but he would not drag his former friend with him. In one of the few incidents in which Nicholas and Gregory had operated more or less together, Nicholas would have to admit later that he had not been as adept as Gregory in dealing with the monk.

---

24. "Pokhozhdeniia Iliodor," *Byloe* (1924), 192.
25. "Perepiska N.A. Romanova i P.A. Stolypina," *KA*, XXX (1928), 85.

Back in Russia, Gregory was dismayed to find he still was not wanted at court. The passing of time had ground down some of Nicholas's resistance; the words of Alix and Anna had softened him, and he was affected by the ingenious and interesting letters of Gregory, who portrayed himself as one who had been cleansed by his pilgrimage and by his long vigils at the Holy Sepulchre. What he needed to complete the picture was a triumphal return to the Volga region where he would be received by large, jubilant crowds. He wanted Iliodor to set the stage for a repetition of the incidents of November and December, 1909. Only Iliodor could do this for him, since Hermogen by this time thoroughly disapproved of Rasputin. Iliodor's own minor victory made him feel independent. He sailed down the Volga denouncing the Duma and calling for pogroms against Jews and revolutionaries. In the Duma, deputies expressed the fear that he might stir up mass rebellion with his processions and heated oratory.[26] Intoxicated with success, the unstable Iliodor, like a man determined to destroy himself, brushed aside Gregory's requests. The *starets* then sent off an angry telegram to Tsarskoe Selo:

> Darling Papa and Mama. Iliodorushka is spoiled. He doesn't obey. Take your time about the *mitre* [promotion] for him. Let it go. We'll see later. He would be all right but he obeys Hermogen. We must be careful.

Once again Nicholas was alarmed by the monk's activity so he began to move against him. Vyrubova telegraphed that he had given in to Gregory's demands, which was not exactly true.[27]

Throughout the summer of 1911 Gregory sulked helplessly in Pokrovskoe. Only the slow passage of time could make the tsar forget his transgressions. In the meanwhile the empress was working for him. She told her husband that the killing of Stolypin at Kiev in September, 1911 may have been a revelation of God's will; according to her Stolypin had malevolently persecuted the *starets*. Remarks such as this weakened the tsar's determination to keep Gregory from court. But his return to grace did not yet come. During the summer and autumn he spent some time in Kiev where the Nationalist deputy from Kiev, V.V. Shulgin reported that he was seen in the company of various conservative politicians. After the assassination, Shilgin added to the mystique of Gregory by saying that

---

26. Gosudarstvennaia Duma. *Stenograficheskie otchety* (St. Petersburg, 1906-17), *Prilozheniia*, III, ii, 59.

27. Trufanoff, *Monk*, 140.

the *starets*, on seeing the premier at a distance, pronounced that he was about to be killed.[28] Some suspected this meant Gregory was part of a plot. His followers remarked it was just another example of his power to see the future. However, it was not a very impressive act of clairvoyance since Stolypin seems to have expected to be murdered at almost any moment.

At the end of 1911 Gregory made a tentative appearance at court. By that time his past sins were forgotten. But he was soon involved in more controversy.

When Stolypin was murdered, Lukianov submitted his resignation at once and it was gladly accepted. Since the spring of 1911 the emperor had been planning to replace both men. In his estimation the Holy Synod was a house that was not in good order. In it Lukianov did not act like the eye of the emperor. As long as questions before it did not directly involve political matters, he was inclined to let the bishops have a considerable voice in making policy. This was consistent with the views of the premier. However, the emperor wanted the Holy Synod to keep peace in the church. For this task he believed that it would need a leader who would stand up to the bishops and concede them nothing. The emperor could not believe that a lenient policy in dealing with the hierarchy might be a matter of policy and conviction. To him it always seemed to be the result of cowardice or lack of will. There were many rumors abroad that in 1913 the tsar would convene a *sobor*. The imagined imminence of this event was causing all sorts of dreams to arise in the church.[29] To resist the angry demands that frustrated hopes were bound to produce, the tsar needed a strong man, not one eternally worried about public opinion.

Gregory was in no position to play a role in the nomination of the Over Procurator, although because of his recent involvement in church affairs he had important interests at stake. His following among the clergy was made up of a heterogeneous assortment of unimportant provincial priests. His presence was opposed by the Metropolitan of St. Petersburg, the monks of the Lavra, and a number of important members of the hierarchy such as the bishops of Kazan and Yaroslavl. Iliodor and his retinue were passing from indifference to hostility. Hermogen had already gone into opposition and was waiting for an opportune moment to strike.

Nicholas chose Vladimir Sabler as Over Procurator. Sabler was a fully-experienced bureaucrat, senator, and member of the Council of

---

28. GDSO, III, v, 55-56.
29. Kokovtsev and Beletsky accepted this charge but offered no proof of it. Nevertheless, the remarks were widely cited by others as evidence that Sabler was an appointee

State. He began his career in the Holy Synod in 1881 and had been Assistant Over Procurator since 1892, when he began training under Pobedonostsev. He had worked in at least three sections of the Holy Synod, and he knew well how the organization operated. On the basis of his training, experience, and his well-known toughness he was well-suited for the new post; by seniority alone was the natural choice for it.

The relationship of Sabler with Gregory was enigmatic, but from the start political gossip claimed that he was the pawn of Gregory, who appointed and controlled him.[30] The two men never met before the end of 1912.[31] Later they met occasionally in salons, but there were no private consultations between them. Sabler held office until 1915, and before he stepped down he acquiesced in several of Rasputin's projects. In spite of his cooperation, Sabler was not always friendly, and the picture of their relationship is a confused one with some evident contradictions. Late in 1913, for instance, the organ of the Holy Synod, *The Bell*, published laudatory articles on Rasputin, reporting that he had played an important role in keeping Russia out of war during a Balkan diplomatic crisis in 1912.[32] This statement was a falsehood. Both the tsar and the Foreign Minister S.D. Sazonov were determined to keep the country out of such a war, and Sazonov was under orders to make this the goal of the country's foreign policy. The policy was highly unpopular in some patriotic circles. The statement in the journal, therefore, enraged the war party. As a result, the conservative press indulged in a spate of grumbling about Rasputin's meddling. This reaction was strong and it could not have been unforeseen. Copies of Gregory's letters written during his pilgrimage of 1911 were circulated in manuscript and had caused some unfavorable comments, because he visited Constantinople and did not mention the need for Russia eventually to annex it. The article in *The Bell* hurt his reputation in conservative circles. Sabler must have been able to predict the reaction, which would have been typical of the pattern of Sabler's treatment

---

of Rasputin.

30. Rodzianko, p. 72.

31. GDSO, IV, ii, 1347. There is no evidence that Rasputin played a role in this or any other foreign policy matter. The emperor and Sazonov worked closely on the problem of war with Austria. Sazonov was under strict orders to avoid war; therefore credit for peace should go to the emperor. Spiridovich in *Raspoutine*, p. 204 reported several other papers besides *Kolokol* mentioning Rasputin's supposed role in peace keeping: *Petersburgskaia gazeta* and *Syn otetchestva*. Rasputin said at this time it was the business of Christians to love and not to fight. See V.P. Semennikov, *Politika Romanovykh nakanune revoliutsii. Ot Antanty k Germanii* (Moscow, 1926), p. 228.

32. GDSO, III, v, 186-87.

of Rasputin: in one way he flattered him and in another threatened him. In 1915 when Sabler was fired, Gregory expressed no regret. He disliked only one thing about the change. Sabler's successor was Alexander Samarin, a bitter foe of Gregory.

Generally, Sabler preferred that Rasputin's presence be kept a secret and that it not be permitted to become a matter of public discussion. The tsar wanted it that way. The silence created the impression that the Over Procurator was trying to protect Rasputin, but actually this was only his attempt to follow orders. Nevertheless, in the meetings of the Council of Ministers he was willing to discuss measures to be taken to bring about the banishment of the *starets*. In summary, the record is not clear enough to permit a firm definition of the relationship of the two. Sabler's ambiguous attitude toward Rasputin, his willingness to give in to certain of his demands and to join with his enemies at other times, suggests the clever opportunism of a man trained to bureaucratic realism in church politics, possibly the most difficult and unrewarding place for an administrator to work.

One thing is certain: only gossip unsupported by facts attributed this appointment to Rasputin.

The emperor ordered Sabler to establish firm control over the Holy Synod. It was first of all important to teach the clerical members that they had to listen to the imperial will. They had been hinting that the church law and ancient tradition might be more important than the desires of the emperor. This notion had been at the heart of the controversy over the canonization of St. Seraphim and had appeared in several other arguments between the emperor and the church. Sabler's first task was to reduce the bishops to obedience, to show them, as Pobedonostsev had, that they were servants of the tsar, and that ultimately it was his power they defied when they rejected petitions submitted by the Over Procurator.[33] It was their duty to listen only to the tsar and not look to important members of the hierarchy for leadership.

We do not know if Sabler's first act in order of business before the Holy Synod happened by chance or design, but soon after taking up his new duties he requested through Bishop Nikanor that a certain Archimandrite Varnava be given the more or less titular post of Bishop of Kargopol. Varnava's name was submitted by a bishop who was a servant of the court, who had no connections with the influential prelates either in the Synod or outside it. There was no expressed doctrine to cover

---

33. "V tserkovnykh krugakh pered revoliutsii," *KA*, XXXI (1928), 211-12.

such a matter, but there was a growing feeling among some of the elements of the hierarchy that they should be in some way consulted about nominations to their ranks. They were trying to establish a principle of cooptation. The proposal concerning Varnava, encountering this small but growing body of opinion, was bound to raise protests. Metropolitan Anthony of Volhynia, the watchdog of hierarchical power, wrote letters against the appointment, circulating them among his friends and stating that Varnava was a creature of Rasputin, who was a *khlyst*.[34] Resistance very quickly developed. Several of the synodal bishops assured Count Uvarov, a Constitutional Democrat deputy, that under no circumstances would they approve of Varnava's appointment.[35]

The archimandrite was a crony of Rasputin; neither he nor Rasputin denied it. He was a peasant from Archangel and as a youth had worked as a gardener. Poorly educated, he expressed himself in the vernacular. His language made him popular with the people, despite rumors that he was corrupt, but the educated clergy looked on him with disapproval. Aside from his closeness to Rasputin, his humble origins and lack of polish were not drawbacks that by traditional criteria would have caused him to be rejected as a candidate for the position of bishop. In the Russian church, simplicity and piousness are considered acceptable substitutes for erudition. A bishop is not the theological arbiter of his diocese, and a number of bishops won fame for virtues that had nothing to do with scholarly pursuits, oratorical skill, or even administrative ability. In the case of the resistance to Varnava it was most likely that the real and hidden issue involved church-state relations. But it was over the question of his connection to Rasputin that the fight became heated. The reasons are not hard to find. The other charges lacked the emotional overtones and popular attractiveness of the Rasputin charge and it was more embarrassing for the government to have to fight on this ground. In addition, the original charge could not be used by the opposition since it might appear to be subversive.

Most likely, Sabler's tactic was to draw out the opponents, to induce the bishops to take a position of extreme defiance. For his purposes the injection of Rasputin's name into the fight was welcomed with appreciation because this would surely arouse resistance. The opposition tried to intimidate Sabler by spreading the story that he was an appointee of the

---

34. GDSO, III, v, 704-5.
35. N.I. Astrov and P. Gronsky, *The War and the Russian Government* (New Haven: Yale University Press, 1929), pp. 17-19.

starets and was now doing the bidding of his master. In the Duma, Count Uvarov tried to help by denouncing Rasputin from rostrum. The Over Procurator countered such moves by posing as the soul of sweet reasonableness. He stepped down from the chairmanship of the Holy Synod temporarily, saying that he was willing to have his assistant, P.S. Damansky, lead discussions. This tactic annoyed and emboldened the resistors, for Damansky was friendly to Rasputin but without any real power. With the Assistant Over Procurator in charge of proceedings, the members were encouraged to express stronger views. Their resolve was further strengthened by the arrival of Hermogen in St. Petersburg in 1911. He made several public denunciations of the appointment on the grounds that it showed the extent of Rasputin's power in the church.

When the clerics were thoroughly committed to all-out resistance, Sabler once again resumed chairmanship of the meetings. He turned to the question of Varnava's sponsors and denied any interference by Rasputin, warning the members that the request for the bishopric came from "the highest authorities," i.e... the emperor. It was obvious that Varnava had more than the support of Rasputin; he was just the sort of earthy peasant-priest who could please the emperor and empress, who in her letters later defended him.

In keeping with tradition, Nicholas was not supposed to be mentioned as the source of such a direct request. The Holy Synod according to the fiction, was supposed to be considering matters of church welfare free of pressure of the government. When Sabler revealed who was behind the promotion of Varnava he was issuing an unconcealed threat. The bishops had taken a strong position of resistance, thinking that Rasputin was to be the foe; now they found themselves in danger of being accused of defiance of the tsar. All of them grew fearful except the doughty Hermogen, who taunted the others by asking if they would bow down to Rasputin. When he substituted the name of the *starets* for that of the emperor, Hermogen made the crisis worse because the accusation became *lèse-majesté*. In the capital there was a rising chorus of abuse directed against Rasputin. When Iliodor came to the city weeks later he decided that the time had come for him to abandon the *starets*, who did not seem to have many friends left in the church. But Iliodor did not have to stand up to the possible displeasure of the tsar. This fact proved too much for the bishops, trained bureaucrats that they were, and they gave in. Hermogen's desperate attempt to make Rasputin the only issue, and thus to hold the line of resistance, failed. Sabler, supported by his statement that he would resign if the members did not approve the promotion, caused resistance to collapse. Within a short time the six bishops of the Holy Synod became docile.[36]

Hermogen tried to lead and rally the opposition. He stressed the theme of Rasputin's influence in the church. He made speeches, held meetings at his residence, and wrote to the press. He insisted that all of this was not disloyalty. "I am ready to obey the emperor, but Grishka Rasputin—never."[37] His utterances neither rallied the bishops nor disheartened the government. The struggle of Hermogen was depicted by the enemies of Rasputin as a crusade to save the church. To anyone who would listen, he would point out that the *starets* was a *khlyst* who sought to control the church. For evidence he pointed to some facts he had learned about Rasputin's conduct in Saratov. He claimed that the government was firm in the promotion of Varnava because Rasputin demanded action and Sabler, his willing tool, was carrying out his order.

The emperor regarded Hermogen as a good man who had temporarily lost his senses.[38] The only issue was the bishop's consistent refusal to accept discipline or to show respect for the orders of the Holy Synod. Nicholas resented the way in which he had deliberately sought to cause trouble by raising all sorts of additional issues. In fact, when Hermogen saw opposition crumbling he tried to forestall the end by bringing up for discussion matters he formerly thought might be best left alone if the primary goal, the destruction of Rasputin's influence, could be won. For instance, he had preferred not to discuss the petition of the Grand Duchess Elizabeth, sister of the empress, for permission to create an order of deaconess. Her project was causing some alarm among the clergy; they thought she was interested in bringing women's religious orders from France, which had recently banned them.[39] In Russia nuns were not organized in such groups. Hermogen chose this moment to resist the plan, and he announced that orders of deaconesses had been forbidden by one of the early church councils. To add to his reputation as the defender of Orthodoxy, he announced his opposition to another proposal that called for prayers to be offered for persons who had died outside the church.

Nicholas expressed his approval of opposition to the proposal of his sister-in-law, but the behavior of the bishop and his public pronouncements had taken on a pattern which indicated that he was baiting the government and was trying to deprive it of control of some church

---

36. Spiridovitch, *Raspoutine*, p. 163.
37. Rodzianko, p. 21.
38. Adam, pp. 207-8.
39. Trufanoff, *Monk*, pp. 216 et seq.

affairs. The Holy Synod was provoked into advising him that the tsar was losing patience. Hermogen rejected the advice and warning, repeating his old charge that Rasputin was in control of the Holy Synod and was using it to carry out his orders.

At this point the church had no desire to go on; it disappeared from the fray. Its leading personages were not accustomed to being embroiled in such controversies, and a few setbacks weakened their resolution. They accepted the deplorable truth that Rasputin was a man of influence. However, Hermogen and Iliodor, who believed in action as well as talk, had originally been attracted to Rasputin because he freely used the name of the tsar, and in a land where the tsar was the head of the church such a friendship could be useful. Not that the two clerics were absolutely loyal to the throne. Each had reservations; they were members of the Black Hundreds, which were becoming more and more dissatisfied with Nicholas, but he was the only tsar they had, and if he could be influenced through an intermediary then some of his faults might be corrected.

In December of 1911 Hermogen sought to remove the evil from court and to end the Rasputin affair with a last desperate act that eventually brought about his ruin. In the presence of Iliodor and several other persons he had an interview with Rasputin that had been arranged by the Minister of Justice, I.G. Shcheglovitov, and another official of the ministry, A.A. Kvostov. The *starets* did not feel sufficiently entrenched at court at this time to be able to survive the big scandal which seemed to be brewing. He had been called at the beginning of December to see the tsar in the Crimea for a brief interview. Rasputin was told that he had not completely returned to grace and was warned to be on his best behavior. As a result, he was eager for peace with his enemies. At the meeting he was respectful of the clergy—as he always was—and when the bishop hurled charges and denunciations at him he was properly contrite. So willing was he for rapprochement that he admitted several of the charges of misconduct with women. He thought that his admission—plus a promise of reform—was all he had to concede in order to obtain peace with his enemies or at least to silence Hermogen during the time of maximum danger. As the encounter moved to its climax Hermogen thought he was achieving his goal of breaking the *starets* by showing him his own iniquities. He lost all prudence and demanded complete submission. When this was refused a fight took place; some later said that Hermogen, a fierce man even in placid times, lost his temper completely and struck Rasputin with the episcopal cross. The *starets* ran into the street, vowing vengeance. But the next day, fearing the reaction of the tsar to the news of the unedifying brawl, he offered peace to Hermogen. Sabler also made a last minute attempt to patch up the affair before the public heard of it, but the bishop spurned these offers and sent news of the incident to the press. In order to defend himself, Rasputin at once moved to the attack by sending Vyrubova numerous telegrams explaining his version of what had happened.[40] Nicholas, in the Crimea, heard this account, and he was also able to read the reports of Sabler, Shcheglovitov and Khvostov, which corroborated the version of Rasputin and contradicted the colorful account of the bishop. This was the end of Hermogen. Shortly after this he went into exile.

Throughout this furious action Iliodor was vigorously fighting and preaching in defense of Hermogen. His activity, however, did much to discredit the cause of the bishop and indirectly to help the cause of Rasputin, the target of his jeremiads. His transition from being a supporter of Rasputin to an arch-foe was too sudden to sound convincing; it had all the marks of too much subtlety and calculation. In the beginning of December the two were still on speaking terms although Rasputin had neither forgotten nor forgiven Iliodor's lack of cooperation during the preceding summer. Within a few days after he announced his break with his friend, Iliodor visited Shcheglovitov, to whom he proposed an assassination plot to do away with Rasputin. The Minister of Justice dismissed the scheme with the dry remark that it was illegal.[41] The emperor was informed of the plan.

Probably Iliodor's most noteworthy contribution to the tempestuous affairs of the year was his blunder leading to the incident that most infuriated the emperor and disgraced the cause of Rasputin's foes. He took the packet of letters Rasputin had given him—six in all—and sold them.[42] Copies at once began to circulate in the city. The words of affection expressed in them were widely misinterpreted, partly because of certain interpolations volunteered by Iliodor who later admitted that the sale of the letters had netted him a large sum of money. Guchov and other Duma leaders examined the originals and pronounced them genuine, and stated that they were harmless expressions of piety containing excessively lofty religious sentiments directed to Rasputin.[43] These remarks made the letters famous and created a clamor from people who wanted to see them.

---

40. *PTR*, Shcheglovitov, III, 130. 41. Kokovtsev, pp. 299-300.
42. Oldenburg, II, 89.
43. Rodzianko, p. 34.

Inevitably, this situation was exploited, and false, pornographic versions appeared.[44] At this point began the rumor that Alix was the mistress of Rasputin and the tsarevich was really his son. Thus, it was believed, a new Rasputin dynasty had been established. Since few could see the originals of the letters, almost no one knew if they were seeing bonafide copies or sensational forgeries. The press contributed to the confusion by making amused commentary on them.

Only if the government secured the originals would the talk stop. Almost at once Nicholas demanded that various officials get back the letters. Because speed was essential the police were delegated this task. Their crude and threatening tactics aroused the suspicion of the Duma, which refused to cooperate until it was certain that the police had no ulterior motives. Next, A.A. Makarov, Minister of Interior, began to hunt for the owner. After months of negotiations he finally was able to purchase the letters. By this time the entire incident had served to convince the emperor that Iliodor, Hermogen and their friends, perhaps even the Duma, were all deeply involved in the scandalous matter. This included the forgeries. Nor did Rasputin escape: his blunder in giving away the letters in the first place came back to haunt him. He tried to escape blame by insisting that Iliodor stole them. Iliodor's wild behavior helped make this story credible. A man who could propose murder would surely not hesitate to steal.

The sordid incidents left the emperor in a mood of anger and disgust. Iliodor and Hermogen were obviously guilty of outrageous acts, but the silent wrath of the emperor was turning against the Duma and his own ministers.[45] He had pleaded with Makarov to muzzle the press; then he ordered him. But the press could not be silenced because of certain freedoms granted by the Fundamental Laws. The tsar regarded this argument as irrelevant and cowardly. Makarov took to delaying tactics, hoping the tsar's anger would subside, but this only convinced Nicholas all the more that his ministers were spineless and were trying to deceive him. By this time, the question of Rasputin had faded from sight.

Before the next round in the battle against Rasputin began, there was a pause while the attackers considered with some amazement what had happened to them. At court leading officials brooded over the painful effects of the events that had taken place since October. They concluded that the monarchy had suffered gravely from the fiascos. It was obvious that the church could accomplish no more in grappling with the

---

44. Kokovtsev, p. 290.
45. Curtiss, p. 374.

Rasputin case; other fighters had to step into the breach if he were to be driven out. Church leaders such as V.N. Lvov drew only one conclusion from the disgraceful Hermogen affair: a *sobor* had to meet soon to free the church from both Rasputin and the state. They made the appeal from the floor of the Duma. On the other hand, the courtiers were not much deceived about the nature of his power. They believed that he could use some influence to sway small affairs; but they did not believe that he was a power behind the throne; they were too close to the throne to be completely deceived on this matter. What concerned them most of all was the loss of prestige suffered by the monarchy due to the presence of Rasputin. Dediulin, the leader of the court faction, understood that in the long run this fact might have important political consequences.

The new allies thought that although Rasputin's power was not great, he was at the same time deeply entrenched. They understood that his tenaciousness had something to do with his friendship with Alexandra, but only Baron Fredericks knew that Alexis was suffering from hemophilia, and only he understood approximately what kind of tragic situation this created. Even Fredericks, who was growing senile, did not understand how Rasputin fitted into the empress's notion that Rasputin was a Friend, in the special religious and ideological way she used that term. Since their analysis revealed that Rasputin was more deeply entrenched than they formerly suspected, they believed that he had to be driven out by a more concerted effort on the part of persons representing a wider range of interests than that which had attacked him in the recent past. All participants had to be absolutely devoted to the monarchy. An alliance was formed among Dediulin, Drenteln, Fredericks, Naryshkina, Prince Orlov, Admiral Nilov, Spiridovich, A.A. Mosolov, head of the court Chancellery, and Count Benckendorf, Chief Marshal of the Court, and several other figures. All occupied high and important posts and several were close personal friends of the emperor, or were intimately connected with him in day-to-day affairs. They were determined to use the old tactic of warning the tsar directly, but they were prepared to visit him in relays, hoping that all of the warnings would have a cumulative effect to eventually accomplish their goal.

To make the presentation more impressive the courtiers asked certain important members of the government to join them. M.V. Rodzianko, president of the Duma, an unusual figure since he was a courtier and member of the Octobrist party; Shcheglovitov, Sabler and Makarov were all drawn in. V.C. Lvov and Guchkov promised cooperation from the Duma; at first they were not acting primarily as deputies but as representatives of prominent and influential segments of the upper classes. The plan was to show the tsar that some of his most respectable and trusted courtiers and officials were speaking with one voice when they warned him of the gravity of the Rasputin affair. Each of these figures had arrived at this point on their own.

In December, 1911 when the tsar returned from the Crimea the political atmosphere had grown warm. The press, especially the opposition publications, *Speech* and *Russian Word*, had for some time been making allusions to Rasputin.[46] The stories were vague and had few facts, but the indefiniteness of the words titillated the imagination of the readers. Even the conservative and widely read *New Times* felt compelled to mention the affair because its publisher, Alexei Suvorin, sensed a good story, one his subscribers were especially interested in because they were monarchists. Even pleas by ministers could not persuade him to stop mentioning Rasputin.

From now on his name seldom disappeared from the papers for any length of time. Whenever he became a center of controversy the journals of the capital were able to add to the air of excitement and crisis by bringing in his name. At the end of 1911 mass opinion did not exist in Russia—not yet. However, the Rasputin affair helped to develop it quickly in the next few years. In fact, it became a *cause célèbre* that burst upon the public from time to time, giving rise to strong feelings and opinions, dividing men along party lines. At first it was discussed among a few informed and concerned people; from them it spread out in society to other groups; then the all-important element of a fast-reacting press dramatized it, seizing the essential ingredients that made it exciting to all kinds of people: sex, religion, politics and scandal in high places. Finally, the politicians entered the picture when they realized that the case could be exploited to their own purposes.

It was at this stage that the Duma entered the picture in early 1912. Rasputin soon became a favorite topic of lobby conversation. Then his existence was mentioned from the tribune where he was described as "the dark forces behind the throne." This became a political code name for Rasputin.

The tsar demanded that V.N. Kokovtsev, the successor of Stolypin, create a law that would permit the government to forbid the press to discuss Rasputin. The premier informed him that the constitution made such a law impossible. Nicholas then turned to the minister of Interior,

---

46. Rodzianko, p. 30.

Alexander Makarov and repeated his demand, claiming that he had approached Stolypin on this subject more than a year earlier. He did not add that Stolypin had sternly rejected the request and had returned the tsar's letter containing it. Makarov refused to sponsor the law. But Nicholas was determined. By late January the Hermogen-Rasputin fracas was widely known and the tsar feared that the press might make much of it, so at a dinner honoring King Peter of Montenegro the tsar took Makarov aside and vigorously insisted on a press law. Makarov felt this was not the right time to bring up such a topic: the Duma was insisting that the wrong man had been banished when Hermongen had been packed off. They wanted Rasputin exiled. On the next day, the hard-pressed ministers, Sabler, Kokovtsev and Makarov made several resolutions: they agreed Rasputin was a great danger to the regime, that the tsar was not completely aware of this, and that in order to end his lobbying for a press law they would have to make a firm, collective statement. They decided they would go to him personally and ask that he get rid of Rasputin. Then, they hoped, Nicholas would drop the politically explosive demand for a gag on the newspapers. At this meeting Kokovtsev, who believed the popular idea that Sabler was a creature of the *starets*, was pleasantly surprised to find the Over Procurator thoroughly in agreement with the idea that Rasputin endangered the monarchy.

The trio got Baron Fredericks to make an appeal first. He had a long talk with Nicholas, who listened patiently to the presentation but rejected it. Makarov and Kokovtsev tried in turn but failed. To the premier Nicholas said of the subject, "It pains me extremely." In his visit Sabler stressed the bad reputation of the *starets* and the way in which gossip was able to exploit it. Nicholas suggested that Sabler get the documents on Rasputin from the Tobolsk Consistory which preserved the materials from the investigation made after the reports of Fr. Peter of Pokrovskoe. These were to be given to Michael Rodzianko, president of the Duma, who would tell the tsar about their contents. Nicholas knew all about them; he was merely trying to get the anti-Rasputinites to look at the facts in the records.

In mid-February the powerful Budget Committee of the Duma discussed the appropriations for the Holy Synod. In its ranks this committee had some of the Duma's most distinguished men: Guchkov, Vladimir Lvov, a leading Nationalist (conservative) deputy, Paul Miliukov, leader of the Constitutional Democrats (liberals), Sergei Shidlovsky, a moderate Octobrist and vice-president of the Duma. In the private debates Rasputin's name was mentioned—always unfavorably—by men of all shades of opinion. The press inevitably took up the cry. For some time the Duma had been pleading with the members of the Council of Ministers to cooperate in maintaining a dignified atmosphere, one free of controversy, while a delegation of British parliamentarians was visiting Russia. Now several Duma leaders broke this tacit agreement. The tsar's contempt for them grew stronger.

Kokovtsev talked to the Dowager Empress, Maria Fedorovna, on 12 February. He pleaded for an hour and a half with her to use her influence to warn her son that Rasputin was a grave menace to the throne. She wept and replied she had said this already but Alix thought Rasputin was a man sent by God and she refused to part with him.

Rasputin knew that Kokovtsev would next go to the tsar, and he knew of the tsar's growing anger. He recalled how hard it had been for him to make a comeback after the long period it had been for him to make a comeback after the long period of disgrace in 1911. He did not want to go on such long travels again; he might never be able to return. One more denunciation might destroy him, so he decided to either head off the interview between the premier and the tsar or to prevent his own name being mentioned.

On the same day as his talk with Maria Fedorovna, Kokovtsev was surprised to receive a note from Rasputin, written in his own hand with his peculiar spelling. "I am thinking of leaving forever and would like to see you so as to exchange some ideas. People talk much of me nowadays. Say when. The address is 12 Kirochnaia, at Sazonov's." On the last day of February the interview took place in the presence of Valery Mamontov, a senator, official in the Ministry of Education, and brother-in-law of the premier. In his writing Kokovtsev was careful to present himself as an enemy of the *starets*, but in the past he may have had some contact with him through Mamontov, and he may not have always been as hostile to the *starets* as he was after he became premier. When he realized Rasputin could be a liability to the head of the Council of Ministers he drew away from him. Mamontov was accustomed to meeting the *starets* occasionally in salons. In 1912 they could not be called friends, but they were still on speaking terms, and although Rasputin's attempts to influence both men in February ended in failure, there were reasons for him to expect success since he was trying to rekindle the spirit of a former acquaintance rather than trying to establish contacts *de novo*.

At this confrontation Rasputin showed off all his usual mannerisms, which his friends assured him would be effective. He marched into the room, seated himself, stared at the floor, then at the ceiling. When asked what he was doing he passed off the question with an inconsequential remark. Then he got down to business: he gazed intensely into the eyes of the premier for a long time, saying not a word. His admirers told him that his physical presence had a hypnotic effect on visitors, and after a few moments of silent exposure to the magnetism of his personality they would be mesmerized. When he looked into a man's eyes he was trying "to observe his soul" as he put it. Many of the important personages who met him reported the same strange conduct in the opening stages of the meeting. Most thought he was trying to hypnotize them. He was, but not in the formal sense of producing a trance. As with the others he was unable to cast a spell. But he proceeded anyway. He said he was willing to promise to leave town if his name were not brought up at the next meeting between the chairman and the tsar. When he was told to go away because he was ruining the imperial family by telling stories that they adored him he protested. "What do I tell? To whom? It's all lies! Slanders!" He yelled, "I don't insist on going to the palace. They call me." Mamontov reminded him of his cheap attempts to peddle influence. He had once claimed he had gotten Sabler his job and had even promised a promotion to Mamontov—in exchange for friendship. After listening to the entreaties mixed with threats Rasputin gave in, apparently. "Very well, I'll go," he said submissively. "But remember they [the tsar and tsaritsa] shouldn't call me back, because I'm so bad and hurt the tsar." At the end of the talk Kokovtsev made a bad mistake: he concluded that Rasputin was a quack and nothing more.

On the next day there was a musicale at Kokovtsev's. His brother-in-law informed him that Rasputin had not yet left, probably fearing he had not accomplished his purpose in the interview and had given his version to the tsar. He said that the ministers were about to deport him by administrative order. This would be usurpation of a right of the tsar; consequently, Nicholas was annoyed. One day later the chairman rushed to court to give his version of what had happened. The tsar got reassurance that there was no deportation planned. Nicholas listened with some gratification, but then Kokovtsev broke the rules of imperial audiences by not permitting the tsar to introduce topics first. He launched into a denunciation of the *starets*. Nicholas listened but he stared out the window the whole time, a gesture every statesman knew was a sign of imperial displeasure.

Mamontov telephoned a few hours later to say that at lunch in the Alexandra Palace Vyrubovna had heard about the talk between the premier and tsar and had told all to Rasputin.

A few days later Rasputin discreetly left St. Petersburg. Some of the courtiers sensed victory, and Benckendorf called Kokovtsov to point out that all three troublemakers—Hermogen, Iliodor and Rasputin—were gone. *Speech* was hailing the premier for carrying out the deportations. Once again Iliodor had to flee like a hedge-priest, sleeping in the open, preaching secretly, and always running from the police. But he was soon caught and shut up in a monastery. Many hoped they had heard the last of both Rasputin and Iliodor.

But the political leaders knew that the tsar had not been won over. They were going to continue their assault, as they did not expect success to come from only a few visits to the tsar. However, before their project could be carried further, unexpected events ruined their plans.

During the controversy involving Hermogen, Sabler and the Holy Synod, Novoselov, a lecturer in theology at the church academy in Moscow and editor of the *Religious and Philosophical Library*, published a pamphlet accusing Rasputin of being a *khlyst*, denouncing his alleged control of the church, calling him "a sly plotter...corrupting of human souls and bodies...brazen deceiver and corrupter...an erotomaniac...a quack." Novoselov's rhetoric thundered on, saying that the bishops were not doing their duty; privately they deplored Rasputin, but while this ravening wolf threatened their flocks they bewailed his presence and did nothing. The essay began with the opening lines of Cicero's first address against Cataline, "*Quo usque tandem abutere patientia nostra?*" ("How much longer are you going to try our patience?"), and it went on to use the threatening and mocking tone of Cicero.[47] The police promptly swooped down on the publisher and confiscated the pamphlet. Excerpts from it appeared in Guchkov's newspaper, *Voice of Moscow*, and at once copies were seized by the police. They also destroyed copies of another conservative paper, *Evening News* which mentioned the pamphlet. The government had overstepped the bounds of its power, and a storm of protest broke out.[48] Since 1906 the Duma had been extremely sensitive on the

---

47. Kokovtsev, pp. 317-19. GDSO, *Prilozheniia*, III, v, 249. These columns contain the reactions of Lvov and Guchkov. Radzwill, *Rasputin*, p. 154 quotes parts of a typical speech of Guchkov that was censored in the Duma reports of 1912 but published shortly after the collapse of the tsarist government. He called Rasputin a "mysterious, enigmatic, tragi-comic figure who seems to have come out of the dark ages," and speculated that Rasputin might be a khlyst or he might be using his religious fanaticism to mask his real role as an ordinary swindler. How did Rasputin, he asked, control or frighten dignitaries of church and state? Not only was he personally evil but also he was surrounded by crooks and adventurers.

48. Rodzianko, p.35

matter of censorship, and whenever the government moved to interfere with freedom of the press the Duma responded quickly and with feeling. The authorities had learned to be careful about arousing the deputies on this issue since few other things could so easily drive together politicians who seldom agreed with one another.

Several things now combined to make the deputies unusually waspish. The electoral campaign was only a few months away—the Duma would end in June—and in spite of the strong protests of the thrifty Kokovtsev, the government was already giving increased subsidies to the reactionary press. The legislature feared that this policy was to be supplemented by a policy of harassment directed against all opposition papers. The silencing of the *Voice of Moscow* was regarded as an opening gun in the campaign. There was a consensus among members of several political groups that whatever was the purpose of the government, there was lurking somewhere behind this move the person of Rasputin, who was trying to protect himself from criticism.

The Duma responded to the threat of censorship and to the supposed intrusion of Rasputin. The budget of the Holy Synod was again to be discussed before committee. Guchkov feared that many of the deputies, who knew that swift and quiet passage of the estimates was much wanted by the emperor, would take the opportunity to retaliate for the confiscation of the journal by making the question of Rasputin's influence in church and government a matter for full-scale debate. Lvov and Guchkov decided that in order to forestall a political crisis the Octobrists would have to lead in a partial airing of the matter; if this were not done there would certainly be a noisy interpellation conducted by less moderate elements. In order to satisfy a majority of moderates the Octobrists had signed a written interpellation and arose to condemn Rasputin.

Speeches followed. Guchkov's was typical. It expressed the deep concern of conservatives and devout monarchists who hated Rasputin because he was befouling the throne and altar, the things they revered. He bitterly denounced those who had failed in their duty to protect the interests of the state and church by remaining silent amidst the scandal of Rasputin's presence. He went on to praise the press for doing what others had feared to do. This was an invitation to editors to begin a campaign of denouncing Rasputin by hammering at the themes of his power, his dabbling in high appointments, and his debauchery.

It was at this moment that the communication between the groups attacking Rasputin broke down. Rodzianko was the key figure in the court-Duma axis, but he was a weak link since he often had an unclear idea of what was happening. To the courtiers, the world of the Duma and parliamentary politics was a dark place about which they knew nothing. It was Rodzianko's job to explain the realities of this world to them. Instead, he offered a perfunctory description of what was happening; he failed to correct their belief that Guchkov was so powerful that he controlled the Duma. To them he was a kind of tsar; they could not imagine how a man could lead a legislature without being all-powerful. As a result of their misconception, the courtiers were inclined to think of the interpellation as a betrayal. In addition to their personal disappointment in the way the Duma behaved, they also felt that when it entered the picture with an aggressive public attack it was time for them to withdraw.

Shortly after Kokovtsev went to Tsarskoe Selo, Rasputin decided to play safe: he fled the city and remained away—except for a few days in March—for several months. Kokovtsev considered ending the assaults. The emperor obviously had heard enough of the entire business and was not disposed to consider the question anymore. Other warnings to him would only run the risk of backfiring, so Kokovtsev sought to discourage other ministers from taking their warnings to the tsar. Despite the *starets'* absence and the reluctance of the premier to talk about him, his case refused to die. Sometimes when it sputtered into life alarms were spread everywhere. For instance, Rodzianko reported that when the imperial family left for Livadia, Rasputin (with the help of Anna) boarded the train, but when Nicholas discovered his presence he had him put off at the next station. A few days later he was in the capital, and he joined a noisy reunion of friends at Golovina's salon. The Constitutional Democrats through *Speech* said he was illegally back in the city but soon would be banished by the ministers. Kokovtsev was alarmed about the tsar's possible reaction to this false report. The wires between Livadia and St. Petersburg came alive as the premier sent coded messages to Baron Fredericks informing him that the articles in *Speech* and the other journals, now in full cry against Rasputin, were not correct. Fredericks replied that the tsar was aware that the ministers had no intention of illegally exiling Rasputin. Police officials warned the *starets* that it might not be wise for him to remain in the city when the tsar was away because feelings were running strong against him and someone might try an assassination. Rasputin himself had asked Kokovtsev, "Well, shall I go? Life has been hard for me here. People make up stories about me." When he asked this question of the police they advised him to leave. He left for Siberia.

At the time when everyone was growing tired of the controversy and many thought it was best not to talk about it, Rodzianko came charging into the fray, full of energy and determination—and confusion. He had seen the tsar twice in January. More than a month went by before they met again. The tsar had carefully hinted to the president of the Duma that he should study the Rasputin records before coming to protest once more. Rodzianko was obsessed with the idea that the Rasputin question was the most important problem Russia faced. He misinterpreted the remarks of the tsar and thought they were an invitation for a full scale investigation. He envisioned himself as opening the eyes of the tsar and saving the country. To support the crusade he tried to put together a coalition of persons who would form a group similar to the alignment of courtiers and officials that Dediulin had tried to lead. His ensemble included the Empress Dowager, a Cossack journalist, Rodionov, Guchkov, Peter Badmaev, a practitioner in Tibetan medicine, Count Sumarkov-Elston, a leading aristocrat and noted eccentric.[49] Others invited to join turned him down; for instance Anthony (Vadkovsky), Metropolitan of St. Petersburg. Kokovtsev watched this motley carnival procession form and then go to its doom.

Before confronting the tsar with the huge dossier, Rodzianko and his wife attended at the Kazan Cathedral in St. Petersburg a special service for divine intercession.[50] He believed that the eyes of the country were on him. It was true that in the halls of the Duma they talked to nothing else that day. But the Council of Ministers looked on the talk with apprehension; Kokovtsev knew that the tsar would regard the interview as an ordeal, and had hinted he would have liked to avoid it.

The president later described with much ingenuousness what happened at the palace. To begin with, he had already committed a serious blunder by giving additional publicity to the Rasputin question when the tsar wanted it buried. At the meeting he also began poorly: he gradually built up to the main topic, in this way increasing the tension under which the tsar had to wait. The last topic mentioned before Rasputin's name was brought up was the bungling administration of Vorontsov-Dashkov, viceroy in the Caucasus. The Vorontsov-Dashkovs were friends of the imperial family and the old viceroy corresponded with Nicholas on a personal basis. The emperor resented Rodzianko's intrusion into administrative matters outside the jurisdiction of the Duma. Then the president moved on and requested leave to talk of something else. The

---

49. Ibid., p. 40.
50. Ibid., p. 44.

emperor paused, steeled himself and gave his consent. Rodzianko poured out his indictment while Nicholas paced up and down behind his desk, nervously chain-smoking cigarettes. Almost at once the emperor, a veteran of such scenes in which so-called documentation was used to prove all the rumors about the *starets*, realized that Rodzianko had let himself be stuffed with all the wildest and most inaccurate gossip. Rodzianko complained that the entire apparatus of government seemed determined to shield this so-called *starets*, who hurt the regime more than did all the revolutionaries. He moved on, making other miscues, and insisted that Hermogen had been replaced only because he had denounced Rasputin. Nicholas countered by saying that Hermogen was a good man but had been guilty of insubordination. Rodzianko said that his dismissal was illegal because a bishop was entitled to a hearing of his peers. This observation was correct, a fact that made the tsar all the more resentful since it was a breach of etiquette to contradict the tsar, and Rodzianko as *Kammerher* [chamberlain] at court should have known this. Nicholas concluded he was dealing with insolence rather than ignorance. But the president droned on, lamenting that Iliodor had been shut up in a monastery because he spoke against Rasputin—a very great oversimplification of the reasons for the monk's fall. The catalogue of Rasputin's victims was ticked off: Tiutcheva, Theofan, Bishop Anthony of Tobolsk, and many innocent girls ruined by the lecherous adventurer. The Holy Synod was his willing instrument, and it silenced or packed off any churchman who crossed the all-powerful *starets*. Finally came the *khlyst* charge. Probably, to get rid of his visitor the emperor said, "I believe you."[51] But Rodzianko wanted more than the statement. He took out pictures and letters from his report. Nicholas replied that he had seen all these inconclusive things when Stolypin had made his attempt to destroy Rasputin. As the meeting ended Nicholas said he could not promise that Rasputin would not return to court.

To the waiting deputies, Rodzianko said that the tsar had received him well. Unfortunately, he mistook a mere descriptive phrase of the tsar—that the report was large—for a value judgment. In other words, the tsar's compliment on its size as taken as praise of its value. Nicholas really thought that it was a foolish rehash of all available gossip on Rasputin. The president seriously misled his colleagues when he told them that as a result of this interview he guessed that the tsar was well-disposed toward the Duma.

---

51. Kokovtsev, p. 317.

Nicholas said he would call back Rodzianko for another talk on the Rasputin case as soon as he had time to work his way through the pile of materials. Through Dediulin he sent word that he wanted the president to personally examine the Tobolsk documents, but he stressed that all this must be kept secret. Rodzianko divulged all. He went to Guchkov, whom by now the tsar personally hated, the Duma members Kamensky and Shubinsky (the latter a Right Octobrist with a bad reputation as a schemer), V.I. Karpov (a member of the Council of Empire), Damansky (the Assistant Over Procurator of the Holy Synod), and J.V. Glinke, head of one of the offices of the Duma, which supplied typists to make copies of the reports. The empress tried to stifle the investigation by going behind her husband's back. She sent Damansky and a court priest, Fr. Vasiliev, to see Rodzianko. She knew that the files contained material describing bad things in Rasputin's past. Damansky said he wanted the files returned. Rodzianko interrogated the official until he confessed that it was the empress who sent him. The priest also admitted that she had sent him with instructions to tell Rodzianko what a pious man Rasputin really was. The president lectured both of them and ordered them out. After this, most informed people knew what Rodzianko was doing. There was no secret.

When the emperor heard of the indiscretions of Rodzianko he resolved not to have him back for another talk. But the president waited to be recalled. When no word came from the palace he sent a note reminding the tsar of his promise. No answer. Annoyed, he went to the premier and pointed out that the president of the Duma had a right to see the tsar; the tsar could not refuse. Kokovtsev thought there must be a misunderstanding, especially since Rodzianko was noted for not always getting facts straight, and he had already misinterpreted the emperor on an important occasion when he thought he was charged to conduct a full investigation of Rasputin. After Rodzianko left the office, Kokovtsev opened a letter from the tsar. Nicholas wrote,

> I do not wish to receive Rodzianko, especially since I saw him but a few days ago. Let him know this. The behavior of the Duma is deeply revolting, especially Guchkov's speech about the Synod estimates. I shall be very glad if my displeasure is made known to these gentlemen. I am tired of always bowing and smiling to them.

The premier was shocked by these words. Of course, the tsar was deeply hurt by the Duma, but his proposed response would destroy all rapport between him and the legislature. The accursed Rasputin business, Kokovstev thought, now threatened months of labor, of delicate maneuvering on the part of the ministers who were trying to get the church and naval budgets through the Duma's hostile chambers. The minister of naval affairs, Ivan Grigorovich and his assistant Capt. Kolchak had prepared painstakingly to go before the Duma chiefs with their case. The formidable Guchkov was in opposition. A temper tantrum now on the part of the tsar would ruin everything. Rodzianko had threatened to resign his court post if he were not heard, and when word of the emperor's affront to the president was broadcast the Duma would surely reject all estimates the ministers asked for. So Kokovtsev rushed to Tsarskoe Selo and outlined to the tsar the facts of the case and some of their consequences. He also reminded him that he could not go back on his word; nor was it constitutionally correct for him to refuse the audience to the President of the Duma; and finally, it was illegal for him to ask the chairman to carry a message to the President. The emperor's pique remained, but he admitted his premier was right, and he promised to send a note asking Rodzianko for a written report and offering to receive him as soon as the imperial family returned from Livadia. Rodzianko was quieted by the response, but his written report contained a demand for the return of Hermogen and the calling of a *sobor*, demands he had no right to make.

In the middle of March Nicholas left the capital. At the railway station there were only three ministers and a few dukes to see him off. The tsar said to Kokovtsev, "You probably envy me, but I don't envy you; I am only sorry for you who have to remain in this bog." With that he turned and walked into the train, not bothering to say goodbye to anyone.

In the next few weeks the air seemed to clear and the political world grew relatively peaceful. With the tsar and Rasputin gone the Duma and the Council of Ministers got on well together. At the end of April, the public turned attention to a brutal massacre in the Lena goldfields of Siberia, where police shot down hundreds of workers. The ministers seemed willing to let the Duma deal with the scandal. The premier remarked with admiration and envy that a young radical lawyer and deputy, Alexander Kerensky (later head of the democratic Provisional Government of 1917 which was overthrown by the Communists) got more accurate news than the ministers could get. Next, the explosive question of the budget estimates of the church schools was brought up in the Duma. Once again the talk turned to Rasputin, Sabler and the Holy Synod. The air resounded with complaints of skullduggery in appointments to ecclesiastical posts. After they had worked themselves up to a high stage of indignation, the deputies again rejected the estimates. Kokovtsev, foreseeing such a happening, and thinking that the Duma was only showing its disapproval, had persuaded the tsar to receive the members at a ceremony marking the last session of the Third Duma, which was about to pass into history. He believed that friendly words from the tsar might persuade the deputies as they returned for the last ballot to change their minds. On hearing the proposition Nicholas made the usual promises—and now in anger made the usual refusal. As the final days of the session neared, the premier reminded him of his promise. Nicholas, seething with fury, snapped back, "I have no time to receive the Duma."[52]

After a long discussion Kokovtsev managed to get him to change his stand, but he warned the premier that the talk about Rasputin had left him with a permanent feeling of rancor for the Duma. "I do not know what I shall tell them; their speeches have been very unpleasant and even revolting to me, and it will be hard for me to resist telling them so." At the confrontation the emperor was correct and aloof. When he came to Guchkov, one of the best known men in the country, he said icily, "I was very distressed with your opposition to the parish schools measure, which is especially dear to me because it was bequeathed to me by my unforgettable father." The deputies withdrew to another room for tea; some were shocked, others fumed at the insult. When they returned to the Duma they once again rejected the estimates.[53]

Rasputin was far away, but his spirit certainly haunted the place where the autocracy encountered the legislature.

§

The essential features of Rasputin's career were developed during the years of 1908-1912: the issues surrounding him, the beliefs about him, and the way people reacted to him. For the most part, the original picture of him implanted in the mind of the public remained unchanged.

At the beginning of the period he was unknown; his name was whispered among palace attendants who observed his occasional visits. A few police were concerned about his connections in St. Petersburg. But after four years his name was constantly mentioned in court, church, society and the Duma, and an attack on him was an attack on Nicholas.

---

52. P.N. Miliukov, *Vospominaniia, 1859-1917* (New York, 1955), II, 105.
53. GDSO, IV, ii, 1345.

The principle issue in connection with Rasputin was his influence. People became accustomed to asking questions about him. They wondered what office he held and asked by what right or in the name of what interest he was exercising his alleged influence. And they asked the most important questions of all: What were his goals? The need for answers to this question was partly the result of the normal human desire to know about the strange and unusual. But for the courtiers and bureaucrats there were greater imperatives. To them answers were important because Rasputin might be dangerous if he was determined to use his influence in the wrong way.

All of the publicized aspects of his career produced a single effect on the public mind. The belief in his power grew as he was reported to be associating, in one way or another, with important figures or to have defeated well known persons in contests of power and place. The repetition of his name in connection with these people left the impression that he was close to the sources of state power, which he was able to use at crucial moments in his own cause.

This view had a certain crude consistency with the facts of the Rasputin case—not as they really existed but as they were imagined to be by most observers of the current scene. In one area—the exchanges of information—the Russian government had changed little since 1905. A gulf still existed between the court and upper levels of the administration on one hand and the legislature and public on the other. Facts lived in a jumble with gossip, rumors and lies. Misinformation and confusion were the result.

From 1908 to 1912 Rasputin had been the protagonist in a number of short but melodramatic conflicts. From all of these he had emerged more or less unscathed, but several statesmen and churchmen had shattered lances in vain attempts to unseat him. Onlookers presumed that in such spectacular clashes in the arena of government only strength could defeat strength. The tsar, Sabler, and the Holy Synod all seemed to be working on his behalf, and for this reason he had been victorious. Such a victor had to have the tsar on his side in any contest of wills and power; an insignificant and unlettered muzhik could survive only if he had the support of the tsar—or if he controlled the tsar. So desperate was the tsar to help Rasputin and to silence opposition that he stripped Hermogen, the most uncompromising foe of the *starets*, of his post in the Holy Synod and packed him off to exile without a trial. In addition to this illegal act, there were his sharp words to the ministers on the Rasputin matter and the plan to suppress the newspapers, both incidents known in a somewhat distorted form among the public. In the summer of 1912 there were rumors that Makarov's days were numbered. Although he was not an enemy of Rasputin, neither was he an admirer, and he refused to defend the *starets* by compromising the government in any way. A few months later he was fired. What were people to think? The rumors about his dismissal were among the first containing speculation on the role Rasputin played in the decline and fall of a minister. In this way his reputation grew because of the roll call of distinguished names that had opposed him in vain. His successful battles were regarded as measures of the kind of power he could wield at court.

The moves of the emperor were interpreted as the desperate acts of a man compromised by association with the starets. It was thought that the emperor was silent and inscrutable because he realized his case did not find support even in conservative circles. His illegal moves were admissions that he could not get rid of Rasputin nor dare to tolerate an opposition that was essentially correct on its guesses about Rasputin.

Nicholas had little interest in either Rasputin or his supposed influence. To him the entire issue was more or less synthetic from the start; the introduction of his name was meant to hide the real issues: the rights of the government in church affairs, the control of the ministers, and the limits of freedom of the press. The emperor knew better than his critics how little influence the *starets* had in either the court or the church. As to the newspapers, law forbid the mentioning of any news about the imperial family unless the publisher got the permission of Baron Fredericks. The papers were ignoring this law, and the emperor probably feared that soon they would ignore others. He surmised that honest observers could see these things; those who did not could only be blinded by evil motives. By this time he had come to regard Rasputin and issues concerning him as two different things. He had a fairly good idea of what the *starets* represented, but he was confused over the meaning of his name when it became a public issue. He might resent the introduction of Rasputin's name into a political controversy, but at the same time he could have his own private reasons for disapproving of Rasputin.

In each battle many factors were involved, only one of these being the desire and interest of Rasputin. His alleged power had actually little to do with his survival of the onslaughts launched by Theofan, Gerasimov, Dediulin, Stolypin, Hermogen, Iliodor, Rodzianko, Tiutcheva, the Duma, and others. His foes often used their own weight against themselves. Rasputin's native craftiness made him do nothing until they began to destroy themselves. In this way he was often as much a pawn of fate as his foes.

There were other things that contributed to the growth of his legend besides his ability to survive attack. He was aware that he could build the public image of himself. In salons and drawing rooms he sometimes hinted and sometimes crudely bragged about his influence at court. His cronies repeated these fantastic claims so that they might help their own causes. His foes also did much to increase his reputation for omnipotence. Not only did they accept his blusterings about his power, but also they embellished these in many ways. There was no claim of his that did not win credence somewhere, and when it became obvious that he was a political liability to the regime, many enemies of the government were willing to join in the game of inflating his reputation. This was a sure way of hurting the tsar. In many ways his enemies were more useful in building his reputation than his friends were. For a long time his friends were nobodies, but his foes were prominent.

Rasputin was never really interviewed, no one took the trouble to ask him what he was doing, what he wanted. If they had asked, he most likely would have been somewhat vague in answering since he did not have clear goals in mind. He did not understand what was happening. Furthermore, there was little planning and plotting by Rasputin in any of his adventures. If defeats of his enemies were milestones of success, it ought to be remembered that he did not initiate any of these actions. He blundered into them and then moved to save himself when he was denounced. For the most part, he was reacting to threats and challenges to his position. Most of the actions were responses to dangers that threatened to destroy him. Reacting to challenges he often faltered in new situations. He advanced in a state of precarious equipoise, on the verge of stumbling or falling but usually getting one foot before him to retain his balance. But at each step his goal was to remain upright, not necessarily to move in any predetermined direction. He knew no better where he was going than did his enemies. And as he became more famous he thought his original faith in his star was vindicated—God must be with him.

Indeed one of the most striking features of this formative stage of Rasputin's career is the contingent nature of the events that led to his success. They often have been regarded as the substance of a tale of triumphs in which he ascended the ladder of power. But there were elements of chance in all this. It is clear that he was not the maker of his own fate. Fortune drove him along. It may be hard to believe that he could have risen from obscurity without conscious design either on his part or on the part of his friends. But even his sponsors envisioned for him a much more limited role than he eventually came to play.

Each step along the way was attended by many hazards for him. His precarious hold on his position can be seen in the unexpected disaster early in 1911 when he had to leave the country. The calamity might have been repeated at any moment during the remaining years of his life. He feared the second time would be the last time. He had prudently tried to insure himself against trouble when Mandryka went on his mission, but the result of his efforts was to temporarily harm his cause. Nor was he able to do much to restore himself to court. As usual, he relied on the faithful attitude of the empress and Vyrubova. But it was the things over which he had no control that eventually brought him back to the palace: the excesses of his enemies, the passing of time which eroded the resistance of the tsar, and the accumulation of distractions that forced the tsar to turn his attention to other matters. It was not until the autumn of 1912 that Rasputin was able to consolidate his position, when the heir was brought back from the threshold of death in what the empress referred to as the miracle of Rasputin's intervention.

Since his foes did not understand how he was able to cling to his position, they often made mistakes by denouncing him at the wrong times with the wrong arguments. Once they were committed to the struggle, they often did and said things that alienated the tsar or tsaritsa for reasons that had little to do with Rasputin. It was injudicious for Rasputin to present the letters of the empress to Iliodor, but it was the monk who foolishly used them in a way which perfectly detract from the cause of Rasputin's opponents and to turn the anger of the tsar against them. It was Iliodor's own foolishness in selling the letters, not Rasputin, that made the tsar his uncompromising enemy. The *starets*' peccadillos, of which Nicholas was aware, were not to be compared with this.

The importance of his career can best be understood when it is placed in the setting of some of the big issues of his day. Among these was opinion in the church where feelings were sensitive and inflamed over matters touching on church-state relations. Clerical leaders reacted strongly to the presence of Rasputin because they feared that he represented some kind of new threat to attempt to reduce the power of the state in their affairs. In the same sense it is easy to understand why the Duma became interested in him since his presence was rumored to have had something to do with the rise of censorship and the threat of imperial interference in the forthcoming elections. Whatever Guchkov thought of the matter, many Octobrists could not resist hitting at the government on the subject of the Holy Synod estimates. Important rights were at stake, The Duma chose this moment to dramatize its power over the state budget. The convenient issue of Rasputin was dragged in as an added ingredient to make the problem of defense of its interests more difficult for the government.

§

The year 1912 marked the beginning of what was for him a quiet period. The year 1913 passed without the emperor calling a *sobor*, and as a result there was rising bitterness in the church. Since the shooting of the workers in the Lena goldfields in the preceding year, labor disturbances and strikes began to increase. Some historians looking back at these years have argued that another revolutionary wave similar to 1905 was building up and eventually, if it had not been checked by the war it might have swept away the autocracy. In early 1913 the leader of the Constitutional Democrats, Paul Miliukov, an intellectual by trade, also thought he detected the kind of increasing political winds that blew before a major storm. In the Duma he decided to take advantage of the church issue so he led his party in an attack on the government, using the Rasputin case perhaps because it drew together by natural attraction a wide range of political opinion, and even had the power to arouse people who normally could not be reached by any genuine political arguments. Thus Miliukov, who personally had no interest in religion, arose to attack the government's meddling, in church affairs. The minutes of the Duma debates indicate much applause for him came from the Left when he called upon the government to free the church, which could then democratize itself. Although the bishops disapproved of his second demand they were gratified with his first. His message was also bound to appeal to the sectarian leaders such as Guchkov. Since the Left was with him anyway he made strong efforts to win support from conservatives. In connection with bad influences in the church, he dragged in Rasputin saying that an ecclesiastical journal credited him with keeping Russia from war with Turkey in 1912. Was this an indication that the *starets* was now beginning to control Russia's foreign policy just as he controlled its church policy? Finally, Rasputin no longer tried to hide his power over the Holy Synod, said Miliukov, and he was shamelessly pushing his cronies into high office. The ignorant Varnava was now nominated for the honored post of bishop of Tobolsk, and Bishop Alexei of Tobolsk, who had for a while been under a cloud of suspicion for heresy, was to be promoted to the position of Exarch of Georgia.[53] For good measure Miliukov threw in an attack on Sabler and included him among the Rasputinites, quoting Iliodor as his authority. Sabler, in his usual way, replied with mildness, protesting that he was not a creature of Rasputin and that the source of Miliukov's remarks was not a good one.

But the storm soon passed and Rasputin once again receded into the background. His contacts with the imperial family were limited. Nicholas still liked him in many ways but found him a bother. For weeks—sometimes for months—Rasputin did not see them. He found it wise not to spend too much time in St. Petersburg. When he stayed there too long the press or politicians began to mention his name, and the longer he was around the more virulent the remarks became. So he was able to avoid much of the notoriety he feared by ducking in and out of the capital and remaining for long spells in Pokrovskoe, Tiumen, or Tobolsk.

Several years passed in this way, and it was not until Russia labored amid the crisis of war and military defeat, with the inevitable troubles that followed, that Rasputin's suspected role in government became a matter for discussion and protest once more.

---

53. Ibid, IV, ii, 1345.

VII

POLITICS, WAR, AND REVOLUTION

Early in June 1914 Rasputin was suddenly back in the headlines. His boasting about his influence at the palace had apparently reached the ears of the emperor. Rasputin decided that it might be good for him to go on another extended stay in Siberia. So he set off on the train, with Vyrubova accompanying him as far as Moscow.[1] On the boat to Pokrovskoe there were other women—one of them, Guseva, said to be a demented former prostitute, followed him home.[2] She was heard to mouth strange religious sentiments from time to time, but no one paid much attention to her because Rasputin was often surrounded by such females. He considered them a nuisance to be tolerated. When drunk he usually cursed and yelled at them to leave him alone, or he gave them tasks; for instance, to go to some sacred tree, usually far away, and to bless themselves beneath its branches, or to go to a holy well—also far away—to drink some of its waters. Although he assigned these tasks perhaps in equal parts jest and anger, his attention only increased the admiration these women had for him. But in Pokrovskoe Guseva caught him unawares near his house and began to scream that he was Satan. Suddenly she stuck a knife in his lower abdomen. The wound was serious, and at first Rasputin's family

---
1. Spiridovitch, *années,* II, 467-69.
2. Rasputin, *Rasputin,* pp. 81, 85.

feared that he might die. He was removed to a hospital in Tiumen where an operation was performed.

Newspapers carried accounts of the attempted assassination and reported that the *starets* might die.[3] On the imperial yacht, Standart, cruising in Finnish waters, attendants noticed a sudden tenseness among the tsar's family. The public speculated about who was behind the attempt. Obviously it was part of a plot. Maria Rasputin, on her way home for the summer after attending school in St. Petersburg, met a journalist on the boat who hinted he was going to Pokrovskoe to cover an attempt on the life of the *starets*. The press discovered that Guseva had been interviewed by Iliodor in the Spring of 1914.[4] There was general agreement that the monk was behind the attempted murder. At court, once again all Rasputin's enemies were lumped together and discredited. The press quickly forgot the matter, its attention drawn to the visits of the French and English fleets to Kronstadt.

Rasputin was tough. To the surprise of all and the chagrin of most he spent the next few weeks in the hospital recovering from the wound. To his friends he sent inscribed pictures of himself sitting under white blankets in the Tiumen hospital. Anna told him of the reactions of the imperial family, assuring him of their deep concern and describing how the empress prayed for his recovery, kneeling in her room before the icon of St. Simeon of Verkhoture that he had given her. A few weeks later the tone of the notes suddenly changed. A new danger had crept into view, stirring fear and anxiety in the palace. There was talk of war. She said that both the tsaritsa and tsar asked him to pray that Russia would not have to fight in the great conflagration threatening the world.[5]

Rasputin understood how his misfortune had aroused great sympathy for him at the palace, and he was aware of how discredited all his enemies were. He also knew how much Nicholas dreaded war. But for some time Rasputin felt he was in semi-exile. His career seemed to be marking time, never having climbed above where it had been before disaster struck in 1911. However, he thought that he might now have an unequalled opportunity to draw closer to Nicholas and Alexandra than ever before. But he misjudged the feelings of Nicholas, who feared war but was determined to fulfill his obligations to France and Britain, regardless of consequences. Rasputin thought that Nicholas was as unalterably opposed to war as he had been in 1912, therefore he decided to

---

3. Rodzianko, p. 104.
4. Trufanoff, *Monk*, pp. 276-81.
5. Vyrubova, p. 105.

take advantage of what he thought was a favorable atmosphere at court. He was right in thinking that the royal pair were in panic and in need of spiritual consolation—his one commodity acceptable to both of them. And he was right in thinking Alexandra was brimming with compassion for him and was sure that only the intervention of God had saved his life. But he knew how dangerous a major conflict might be for Russia. In addition to his personal feelings, he was in contact with politicians who were convinced that Russia needed another decade of peace in order to strengthen its political and economic order before it risked involvement in war again.[6] They feared that the strains of a full-scale European war might be more than the autocracy could stand, that the holocaust of war might sweep away the empire into revolution. Rasputin knew how little the people cared about war. He could visualize how their indifference could in time of privation and death be converted into hatred and violence. Therefore in the mood of a seer, afflicted with a foreboding sense of tragedy, he took it upon himself to step into politics and give direct advice to Nicholas to keep Russia at peace.

> My Friend:
> Once again I repeat: a terrible storm menaces Russia. Woe, suffering without end. It is night. There is not one star. A sea of tears. And how much blood!
> I find no word to tell you more. The terror is infinite. I know that all desire war of you, even the most faithful. They do not see that they rush toward the abyss… You are the Tsar, the Father of our people. Do not let them throw themselves and us into the abyss. Perhaps, we shall conquer Germany, but what will become of Russia? When I think of that, I understand that never has there been so atrocious a martyrdom. Russia drowned in her own blood, suffering and infinite desolation.[7]

---

6. For expressions of this view see: M. de Taube, *La politique russe d'avant-guerre et la fin d'empire des tsars* (Paris, 1928), p. 331; R.R. Rosen, *Forty Years of Russian Diplomacy* (New York: Allen & Unwin, 1922), II, 149.

7. Rasputin, *My Father*, pp. 77. On February 24, 1915 (N.S.) Rasputin expressed the same view to Paléologue. "There are too many dead and wounded, too many widows and orphans, nothing but ruins and tears. Think of all the poor fellows who'll never come back, and remember that each of them has left behind five, six, ten persons who can only weep! And what about those who do come back! What are they like! Legless, armless, blind! It's terrible. For more than twenty years we shall harvest nothing but sorrow on Russian soil!" But he understood that Russia could not make a separate peace. To

He had begun his program of warnings by sending telegrams, but when these had no effect he turned to the more formal words of the letter.[8] Alexander Kerensky learned of the contents of these cables from a fellow Duma deputy, N.S. Sukhanov. Sukhanov had been elected from Tobolsk, and he was able to persuade the postmaster of that city to show him the Rasputin telegrams. One read: "Do not declare war, chuck Nikolashka," (referring to Grand Duke Nicholas Nikolaevich, who looked with favor on war with Germany) and "Declare war and they will be shouting down [with the tsar] again. Evil will come to you and the Czarevich."[9] In the last sentence Gregory was plainly saying that war would doom the dynasty. Thus he joined the small band of prophets who clearly foresaw the worst consequences of the war.

But Nicholas had found another kind of support and consolation. To the surprise of all, Russia at the start of the war experienced a *union sacrée* in which all parties drew together to defend the motherland.[10] Even the revolutionaries, with the exception of a few leaders, joined in. Eventually, the tsar foolishly destroyed this good will, but on the outbreak of the war he was amazed to find himself carried to a summit of popularity that he had never known before. At great public ceremonies in St. Petersburg and Moscow masses of people cheered him; at one memorable rally in the capital a huge crowd knelt before him and sang the anthem, "God Save the Tsar."

The war was national salvation. Nicholas did not think it would be concluded without suffering, but it had healed the wounds inflicted on the nation by internal political fighting, and it provided him with a great crusade in which he could unite his people behind him. For the first time he was a leader of the entire nation, and at one stroke his worst problems seemed to be swept away. He felt all along he had known that the divisions in the country, the threats of revolution, and the fears that Russia was heading for catastrophe were all lies dreamed up by weaklings and fools.

Paléologue he said, "You're right. We must fight on to victory." See Semennikov, *Politika*, p. 228.

8. On hearing of mobilization Rasputin wired a prophetic statement to Nicholas II: "The war must be stopped. War must not be declared: it will be the end of all things." Alexandra in a letter of November 1, 1915 reminded her husband of Rasputin's wisdom: "Our Friend was always against this war, saying that the Balkans were not worth the world to fight about and that Serbia would be as ungrateful as Bulgaria."

9. Alexander Kerensky, *The Crucifixion of Liberty* (New York, 1934), p. 190.

10. *Pravitelstvennyi vestnik*, no. 165, August 9, 1914.

When the telegrams and the letter arrived from Pokrovskoe he rejected them. But he went beyond that. He strongly resented Rasputin's interference in matters of state, and he expressed his feelings freely at the palace.[11] Since these angry remarks contained threats as well as denunciations, Rasputin was frightened. When he recovered from his wounds he crept back cautiously into the capital while all eyes were riveted on the first battles of the war. Until the middle of 1915 he was once again in a position of danger and could at any moment be swept away. He realized what few suspected—how poor were his sources of information at the palace and how easy it was, therefore, for him to guess wrong about what he might dare say or do. General Resin, in charge of all communications at the palace, confided to a friend that Rasputin often placed calls to the palace but he usually talked only to a few friends.[12]

§

At the start of the war Russia reacted hastily to French cries for immediate assistance, and an offensive was launched at East Prussia by the unprepared Russian armies. As a result, they were defeated at the Battle of Tannenberg. Although the victory for Germany was impressive, it was not decisive. Russia's army quickly recovered. However, some parts of public opinion descended into a state of gloom and defeatism and never recovered hope. Grumbling was heard at once, most of it directed against the tsar. But in the remainder of the year Russian armies fought a series of large-scale battles between the Carpathian Mountains and the Baltic Sea. They acquitted themselves well, parrying German and Austrian drives and launching offensives of their own in a wild war of maneuver.

By the end of March, 1915, Russia had smashed its way into the Austrian Empire, occupying positions overlooking the plain of Hungary. After this, Austria seldom showed aggressiveness on the Eastern Front; it was permanently hurt. Germany was forced to take a new look at its strategy in the East. It had been dismayed to find the Russians still fighting hard after seven months of furious battle. The staying power of the tsar's forces was greater than had been imagined; conceivably in 1915 they could ruin Austria and turn the tide of battle against the Central Powers. After a period of arguing, the generals decided to conduct a holding ac-

---
11. Vyrubova, p. 104.
12. Buxhoeveden, p. 143.

tion in the West in 1915 while the main effort was directed against Russia. In France the commander-in-chief, Joffre, nibbled at the Germans while he prepared for an autumn offensive—but by then Russia was fighting for its life. On 2 May the Kaiser's armies, along with the Austrians, hurled themselves at Russia. The Russians took the charge of both empires, held for an instant as their lines bent under the violent assaults, but then broke into full retreat when the enemy tore gaps in the line at Gorlice-Tarnow between the Carpathians and the Upper Vistula. The next five months were agonizing for Russia as its armies reeled back in defeat, losing almost a million men, suffering huge losses in matériel and artillery, and giving up Poland, the most advanced industrial region of the empire.

From the start of the war Russia was a blockaded country. It suffered from the results of decisions made two generations earlier when the government resolved to give a second-place priority to the rail and water communications in the northeastern part of the empire. These could have provided easy access to the Baltic and White seas—the best trade routes to western Europe. The blockade was clamped on the country at the end of October 1914 when Turkey entered the war and cut off the routes of the Black Sea ports. Germany already stood astride the rail lines going west. Vast economic dislocation now descended on the empire. Travellers reported seeing huge reserves of grain and butter standing in the open beside Siberian railroad stations. They could not be moved to the population centers in the European portions of Russia. The armies suffered from a severe munitions crisis that the government seemed unable to deal with. The inadequate railways limped along trying to supply the civilian and military sectors of the economy. The loss of the manufacturing regions caused all kinds of shortages to develop at home and at the front.

The French had proved that the problems of *Materialschlacht* could be solved even when a large and critically important area of the country was in the hands of the enemy. Russia took the same steps that other countries did, hesitantly and at first slowly, but in the end the results were similar although they came too late to save the empire. By early 1917 the armies were being better provisioned and those who were concerned only with the country's military problems were growing optimistic. They thought that a few modest victories in the spring would depress the enemy and hearten the morale of the home front.

But tsarism did not solve the problem of civilian food supply in big cities, and it failed completely to deal realistically with the most critical resource of all: public confidence and morale. The 1915 defeats brought a typical remark from the leading publisher, Boris Suvorin, "I've lost all hope...we're doomed to disaster from now on." This feeling was widespread in the country. In Moscow crowds rioted in front of the Convent of Martha and Mary, which was run by the sister of the empress. The mob reflected a widespread feeling that the dynasty was either treacherous or so stupidly inept that it could not lead the nation safely through the war to victory.

From the opening guns of the war the autocracy revealed its inability to draw together all the country's strength. But it dared not turn to the *zemstvos*, or the many private groups that tried to solve some of the supply problems of the army. In most of the large cities private citizens organized production to make clothing and medical supplies. Although there were some abuses connected with these non-official efforts (the zemstvos, for instance, housed some draft-dodgers, called popularly "zemstvo hussars"), they enjoyed overwhelming public confidence for their record of accomplishment. The government agencies were unusually slow and inefficient, and they struggled in a sea of confusion, always with some hint of corruption about them. Nevertheless, there were reasons why the government could not turn to the public for help.

Members of the liberal opposition had once speculated about the possibility of pushing their cause during a time of national emergency. They believed that such a moment would give them a chance to demonstrate their usefulness to the country, but they had no illusion that tsarism would be reasonable and receptive. Only when it had been driven to the wall in extreme emergency would it accept help. Then, the liberals felt, they would both offer aid and demand a price for it. The price would be their right to participate in the executive part of the government by having some say in the appointment of ministers. To the defenders of the regime this idea was a form of revolution, and the government did not tolerate revolution. As one defender of the old order put it, the opposition intended, under the guise of delivering boots to the army, to carry out social revolution.

Despite the staggering losses, the commander-in-chief, Grand Duke Nicholas Nikolaevich, kept his armies intact. However, his achievement was not appreciated, and the country began to lose hope of victory. The defeats created the conditions the Duma opposition had hoped for. The government had left the country into a catastrophe and did not know how to lead it out, for it possessed no more spiritual or political resources to draw upon. Most people believed that the fate of Russia depended on the solution to the so-called munitions crisis—actually a web of individual problems concerning the production and movement of materials for the war effort. They reasoned that the government could not do the job; therefore, it would have to call on the help of persons in politics and industry. Eventually, only when the tsar found Russia on the edge of destruction did he give in and permit the establishment of the kinds of institutions that had mobilized other European states for war. A series of special councils were created to supervise the rational use of key sectors of the economy.

Although the special councils solved many problems of supply and allocation they could not solve all. Their powers were limited, and they needed some time to acquire experience. But they quickly made an impression. However, as the tsar feared, concessions to the public only increased its desire for more direct participation in running affairs. The foundering of the government and the success of the councils gave fuel to popular agitation for the swift granting of liberal reforms.

The tsar had used the war as an excuse to avoid calling the Duma to meet; the opposition used defeat to begin a campaign to enlarge the Duma's power. They proposed to help the tsar save the country by mounting a campaign in which they demanded a ministry of confidence, as they called it—that is, one enjoying the confidence of the nation. Only the Duma could express this confidence. To prepare for the day when the tsar would at last decide he had to call on the support of the public, there was created inside the Duma a Progressive Bloc—a typical parliamentary coalition of several political parties united to offer themselves as a basis of support for the government in the legislature. The Bloc was made up of center and left-of-center parties and personalities. Their demands were relatively moderate. The tsar may have been tempted at times to accept the hand they proffered. To do this would have solved many of his political problems. But instead he reaffirmed his power. As the country drifted deeper into crisis more members of the legislature began to associate with the Progressive Bloc, which seemed to be growing into a real parliamentary majority and was therefore a threat to the government. The empress, for instance, believed that the real danger to the tsar came not from revolution in the streets but from a political coup involving the Duma, some military chiefs, and certain members of the imperial family.

The defeats of 1915 thus brought the government back to a familiar position; it was 1905 all over again. But this time there was no need for the opposition to make an alliance with the open revolutionaries. There were too many other paths to power open. By playing their cards wisely the opposition might force the tsar to give in; in other words, parliamentary action might compel him to do what demonstrations, street fighting, and peasant riots had largely failed to do in 1905-06. In the summer of 1915 Nicholas revamped his cabinet and seemed to be responding to the voice of the country. He dropped men who were notoriously offensive to the progressive sections of society, and appointed men who were moderates. But this was not the end. Take one more step, the Duma asked, and recognize the Bloc. The tsar realized he was on the verge of an action similar to the granting of the October Manifesto. He still felt that in 1905 bad advice had forced him to go further than was necessary. He would not make the same mistake again. Moreover, he had run out of room, there were no more alternatives. The Duma had the initiative now and it forced a choice on him: either avoid reform and continue the swift slide to doom or give in and be saved. He was aware that more reform would take him across the political line into the dark land where the legislature would be equal to him. He could not move, he thought. Things would have to remain the same. He did not have the right to renounce his coronation oath and forsake his son's birthright. He proposed to wait for God to grant victory after the long ordeal was over. The country would be redeemed, and after the war the people might feel differently about his rule. If they did not like him… well, there was always a way out for all: abdication, when a new tsar with a new oath would promise to obey not the voice of God but the will of the Duma.

§

As the throne began to rock dangerously Alix acted. But her house was in disarray. Vyrubova was so loyal that the empress could occasionally enjoy the pleasure of a spat with her without having to fear betrayal. Sometimes they feuded like fishwives. In the summer of 1914 Vyrubova went away for a while. Except for a brief and almost secret trip to the capital in the autumn, Rasputin was absent for a long time. Would the threesome break up? With his communications link broken, Rasputin had no reliable way of sending or getting word to or from court. At the beginning of 1915 Vyrubova was critically injured in a railway accident that left her a semi-invalid.[13] After this she used crutches for a long time. Before the accident she often made phone calls to Rasputin's flat; sometimes she

---

13. Vyrubova, p. 118.

dropped in to see him. Although she was unlikely to notice the seamier side of his life, there was always a danger to Rasputin in her sudden visits. She often had some court ladies with her, and not all of them could be trusted to keep quiet. This fear kept Gregory alert and his dissipations had to be kept within bounds so that he could quickly put his house in order when a phone call came from the palace attendants to announce that Vyrubova had left Tsarskoe Selo for the short ride to St. Petersburg. As a result of her semi-invalid condition she could no longer make sudden trips to his apartment. Gregory lost the fear of being discovered and became more abandoned in his carousals. All kinds of unsavory adventurers began to loiter about his place, and he did not bother to hide them. But there was one more consequence of the accident. Since he had not been in full favor at court, he did not hear of Vyrubova's injury until the second day. When news was brought to him he jumped up from a meal and rushed to Tsarskoe Selo in a car loaned to him by a friend, Countess Witte. At the palace he found Vyrubova supposedly in a coma crying out for "Father Gregory." Last rites had been given. As he told the story, he withdrew into another room to pray. Then he returned and leaned over her for a moment. Putting his mouth near her ear he called out in a loud voice for her to wake up. She regained consciousness and then began a period of convalescence. Alix considered the episode another example of his miraculous powers.[14] No doubt Nicholas had to listen to this story and nod agreement about its meaning.

In 1915 despite the crippling of Vyrubova and the semi-retirement of Rasputin from court, Alix began to intrude with some consistency in affairs of state. It was one of her husband's few determined acts that caused her first meddling in politics. Nicholas had decided to use military victory to add lustre to the throne, so in the spring of 1915 he announced a plan to tour the recently conquered parts of Austrian Galicia, walking over land that Russian princes had once claimed in the thirteenth century. Responsible ministers and officials warned him that this was a risky venture. One could never be sure of the fortunes of war; if the Central powers recaptured this territory after the tsar's visit then the monarchy would suffer a loss of prestige. The Grand Duke Nicholas had fought the war up to this point according to plan: he was essentially defensive, keeping his lines fluid and trying to avoid having his armies trapped or chewed up in stationary battles. He did not want to be saddled with the responsibility of having to stand and fight in Galicia merely because the

---

14. *PTR*, Beletsky, IV, 501-2. Mosolov, p. 162. Spiridovitch, *Raspoutine*, pp. 17-18.

tsar had been there and had ostentatiously slept in a bed that once belonged to the Emperor Francis Joseph.

Gregory at this time was growing more alarmed over the suffering caused by the war. He rejected anything that might increase the burdens of the people. In doing this he often had to go against the pleas and demands of some of the circle around him. Adventurers, some churchmen, and bureaucrats such as Count Vladimir Bobrinsky and Governor General of Galicia, who dreamed of an enlarged viceroyalty, saw that the occupation of the new lands offered them opportunities for power and riches. Alix feared the results of a possible retreat from the Carpathians, so she hesitated to endorse her husband's proposed trip. But the problem was made difficult for her since she was looking after the interests of the dynasty. To her Rasputin reported about rumors circulating in the capital as early as September 1914, that the Montenegrins and the Grand Duke Nicholas were hinting that the duke, the savior of Poland and the liberator of Galicia, should be given an independent crown in the new lands.[15] Alix could not be sure that the duke did not have even greater dreams—perhaps of a Russian crown some day. The tsar's trip might result in putting himself in checkmate. But it the Central Powers returned—what then? She remained lukewarm in her support of the tour. Nicholas went nevertheless.[16]

But the disasters at the front in May-August 1915 brought her to her feet in alarm. The political actions of the tsar worried her even more. She did not approve his determination to meet the disasters with long-suffering patience. She wanted him to take command of the country, to lead it and drive it ahead. She called on him to be an imperial whip lashing out against the lagging parts of the nation, roaring out orders, threats and reprimands to the cowardly ministers. She found his quietism unbearable, and yet she could do little to make him act. She wanted him to be dynamic, not passive. She professed to admire the turn-the-other-cheek philosophy in others, admiring Rasputin's willingness to suffer slander without making public response. But such conduct was unbecoming in a tsar. From his position her thought moved on. She became, in fact, the perfect virago, the raging wife who sought to dominate her mate. Dr. En-

---

15. A to N, September 20, 1914.

16. Rasputin strongly opposed the trip and sent a telegram to Vyrubova about it. "God will help; but it is too early to go now, he [the tsar] will not observe anything, will not see his people, it is interesting, but better after the war." A to N, April 6, 1915. Buxhoeveden, p. 203, said that Alexandra at first disapproved the trip and later tried to hide this reaction.

causse had taught that women were guardians of the most subtle forces of human nature, that in them resided something that spoke clearly and intuitively, and it was superior to the rational faculties of men.[17] Alix once wrote to the Grand Duchess Victoria that there were times when the weaknesses of the monarch demanded that his consort step in and participate in running the state.[18] In 1915 she must have thought that such a time had come. She could contribute much, more than mere intuition. She had a strong will, and Dr. Encausse had said that the will could be omnipotent, it could actually control the universe when it was bolstered by knowledge that came from God. She believed she had access to this kind of knowledge through Gregory who was a Friend of God, a man sent by the Almighty to provide wisdom to the rulers.

Alix began her period of maximum activity in the autumn of 1915. Nicholas, against the advice of almost everyone, took command of the Russian armies. Alix proposed that she should play the role of the Empress Catherine the Great, an unfortunate allusion since that woman usurped power from her husband and probably was involved in his murder. In September Nicholas wrote from military headquarters, the Stavka, welcoming her assistance. A month later he added:

> Yes, truly you ought to be my eyes and ears there in the capital, while I have to stay here. It rests with you to keep peace and harmony among the ministers—thereby you do a great service to me and to our country. I am so happy to think that you have found at last a worthy occupation! Now I shall naturally be calm, and at least need not worry over internal affairs.[19]

Unfortunately for her, Nicholas, as usual did not live up to his word, and in the future he denied most of these promises.

§

On the eve of World War I Rasputin was a controversial and notorious figure in Russia, but little was known about him. In St. Petersburg his enemies accused him of being a false prophet who deceived gullible

---

17. P. Encausse, p. 498.
18. V.M. Purishkevich, *Dnevnik* (Riga, 1924), p. 77.
19. N to A, September 23, 1916.

courtiers by posing as a holy man. No doubt when he became an object in the public eye he was no longer principally concerned with religion. His old faith was gone; it had never been whole or complete, and it was soon worn away by time and his own betrayals. By 1914 he had in fact travelled a long way from the days of the revolution when he had first come to court. He had lost his innocence and he gained some power, especially in church politics. In his search for more power—and to retain what he already had—he was forced to appease the empress, for him the last stage of the decline of his faith. Alix did not corrupt a good man; he was already bad, but with some redeeming features.

What began as a scandalous comedy turned slowly into a tragedy for Rasputin and for Russia. Before winning his historical reputation he was a man of affairs in a limited sense. He visited several salons and was alone as a *starets*. His advice was sometimes sought, and he was permitted to express conventionally pious views. But even then he had another side to his existence—in the shadows he led a life of quiet dissipation, frolicking in the dens of the capital and in the homes of some well-known manipulators. However, by 1914 this side of his life was no longer a secret for most people. He still appeared from time to time at fashionable salons such as those of the baronesses von Hildebrand and Rosen, and the great dinner soirees of Mademoiselle Ostrogradskaia, but his dissipation was getting out of control, and he was no longer able to hide it at will. Luckily, for him circumstances at the palace and in the city made concealment less necessary.

Rasputin's story from 1905 to 1916 is the familiar tale of the decay of faith. This is a special kind of problem in the psychology of religion. Unfortunately, his experience is mostly hidden from us. He did not write down formal statements to describe his own genesis. Therefore, if someone tried to describe the process he would have to use the scanty data of intelligent but poorly-informed observers and the gossip or hearsay of scandal mongers. The kind of evidence used in an attempt to show the rise of his faith might also help to describe its decline and fall.

We do not know his first reactions to St. Petersburg, but he probably did not find it as St. Augustine found Rome—"a cauldron of unholy loves." This is not the stylized tale which pious folk liked to tell of the innocent rustic who came to the great city where he eventually lost his faith. But there can be no doubt that Gregory's crude outlook was shaken by what he saw and heard in St. Petersburg. Experience forced him to re-examine his original beliefs and to modify them in order to exploit the new situations he encountered in the city. He grew more confused about his own thought. As he became a manipulator and confidence man the forms of faith remained the same, but they had less meaning for him. His views did not square with his life. The conflict and emptiness of his forms made him look foolish and sinister to his foes. He was puzzled as they were about what he was and what he believed, but he never lost faith in the conviction that he was a vessel of truth. Emotion and intuition were his guides, but they were bad; they gave him in the end a religious feeling without religion. His tattered creed grew into a melange of pseudo-religion tinged with a cult of sensuality. His faith gave room to fetishism and social gospel, but Bacchus and Orpheus were among his first gods, and his rite was the drunken party which could begin at any time he bestowed an enthusiastic *abrazo* on the nearest female.

But he never became interested in organizing a cult. He had some hangers-on and a few cronies, but among them there was no sense of corporate fellowship. His apartment was always cluttered by a half-dozen or more female friends who adored him extravagantly as a saint and loitered about, waiting for his blessings. But he either cursed or scorned them, depending on his mood, and they loved him for it. Although he was normal many of his friends were not. Those closest to him were masochistic in their religiosity, in need of reassurance that they were being forgiven for their sins, real and imaginary. Women who were honest sensualists were not attracted to him; they had no reason to hide their lust under a false spiritual garb. Therefore, they were inclined to see him for what he was—an old drunk who had some money and a bit of political influence. They might pursue him, but not because he enchanted them. With them he caroused in the fleshpots of the capital, being observed coming and going from restaurants and bathhouses.

Most of his visitors came in connection with their petty business transactions or with pleas for help in dealing with the bureaucratic apparatus of the state. They usually came to him after many desperate but unsuccessful attempts to petition or interview some high-ranking official. When they heard about the all-powerful *starets* they sought him out, but there is almost no evidence that he was able to help anyone. Nevertheless, the legend of his power endured and they kept coming. Despite stories spread by his retinue, few of the visitors were persons of importance, and no countesses or duchesses threw themselves at his feet in supplication. But to take care of visitors he conducted a kind of floating court that met in his apartment or in a house belonging to a friend, in a hotel dining room or a well-known restaurant such as Donon's or the Yar, where he liked to drink and listen to the gypsy singers.

By 1915 he was a well-known drunken satyr, but his sexual exploits seem to have been considerably overestimated. Although he made love *con brio*, drinking seems to have been his main vice, and like most Russian peasants he got drunk easily on a small amount of liquor. He was never a true hero of the bottle. He drank Madeira, a heavy sherry-like wine; its crushing hangovers left him constantly sick and ill-tempered. Nevertheless, he reveled in long nights, and some of his capers turned into noisy brawls that provided amused or shocked gossip among persons in government and court.

This is the Rasputin who became famous. A graphic record of his decline and fall to this point is preserved in two photographs of him. The first was taken early in the century. He stands erect facing the camera, his eyes looking ahead, one hand folded placidly on his breast and the other raised in a gesture of blessing. His expression is pleasant and appealing. Indeed, most people who met him at this time reported on his attractive personality and his straightforward, laconic Siberian speech. Officials, however, were disturbed by some vague quality in him; he was not humble like most peasants, and he had an unnerving way of coming directly to the point. The second picture—and best known—was taken near the end of his life, probably sometime during the war. It is a close-up of his head. The crafty eyes are almost closed, hidden behind narrow slits. The expression is calculating and almost demonic. His hands thoughtfully finger a few strands of hair in his beard as he peers with ruthless concentration, seeming bent on some terrible act or remark.

Rasputin became a familiar figure in St. Petersburg society. Usually he was seen in a cab, hunched low in the back seat, crowding himself into a corner as if in fear, a wide brimmed hat pulled down low on his forehead. At other times he was on the street, always in the presence of a large number of friends. His appearance was striking as he walked along with cat-like strides, his muscular figure bent slightly forward, seeming to slip through the air. Indoors he wore heavily embroidered silk shirts, red or white being his favorite colors, and around his middle he tied a braided and tasseled cloth belt through which he constantly thrust his flattened hands. Baggy pants were tucked into the expensive black boots that his friends always kept well polished. His long hair parted in the middle and fell on either side of his face, sometimes hiding his features. Toward the end of his life he was full of mannerisms and affectations. He never addressed anyone by name but immediately on meeting a person thought of a nickname. Anna was "Ania" and Witte "Vittia." When these simple names appeared in his telegrams, which were widely circulated and read, they were thought to be part of some kind of secret code. He could carry phoniness to the nth degree: in his speech "o" was given a peculiar sound of his own choosing. When talking about politics he was lofty with pride in alluding to his power at the palace. He liked to mention coyly that he talked to the empress. To a petitioner he would say "About your problem, I'll talk to her," pausing after "her," slyly avoiding identification of who this was but all the more drawing attention to his supposed ability to mention anything to the empress.

We have several records of his typical behavior at this time, kept by a woman who visited him in both St. Petersburg and Moscow on several occasions between March 1915 and May 1916, petitioning him to help her save her mother, German born but a long-time resident of Kiev who was threatened with deportation during the great retreats of 1915.[20] In Moscow when she entered his place he made her sit beside him, but then he ignored her. The room was filled with people, mostly women. After a while he turned and let his glance bore into her eyes. Then he suddenly said one word, "Drink!" Later he handed her a pencil and some pieces of paper. Another word, "Write!" He dictated a rambling note.

> Glory to simple folk, evil to rebels and the wicked. The sun will not warm them. Forgive me, Lord, I am a sinner; I am an earthly being, and my love is earthly. Lord, make some miracles. Humiliate us. We are your creatures. Great is Your love for us. Do not be provoked against us. To my soul send submission and Your divine grace. Save me and help me, Lord.

All present strained to hear each of his words. Throughout the evening he dominated all conversation by suddenly switching topics without warning, sometimes jumping back and forth in conversation with no connections between ideas. At one moment he suddenly talked of war, saying how terrible it was and bragging that if he had not been sick in Tiumen in August 1914, he would have forbidden the emperor to go to war. "What good is it?" he asked. When supper ended everyone went to a sitting room. After a while Gregory jumped to his feet and demanded that one of the women play the piano. Then he began a dance, first alone later in company with one of his female friends whom he invited to join him by making gestures with his hands as he moved about. His visitors marvelled at the grace and lightness of his movements. The truth is that

---

20. Elena Dzhanumova, *Moi vstrechy s Gr. Rasputinym*. (Petrograd, 1923).

he was an extraordinary dancer—even his enemies admitted that. He preferred steps with quick or violent movements; the Cake Walk, a recent importation, was one of his favorites and he executed it with great abandon. However, on this particular evening, he preferred something more sedate—there were strangers present. Later in the evening he found time for a brief discussion about his visitors' request, telling her at last that she should come to him in St. Petersburg where they could talk about the matter at length.

In September, 1915 she visited him in his flat at 64 Gorokhovaia where he had five rooms on the third floor of an unpretentious building. The court had rented this apartment for him early in 1914, and he had moved in with his two daughters who were away most of the day at school. In the evening he usually went out with friends and stayed away until the early hours of the morning when his drunken revels gradually played out. Police agents stationed near the doors and under the stairways watched Rasputin, both to spy on him and to protect him. Inside the flat several female housekeepers directed by the aged ex-nun Akulina tried to preside over the confusion of his household. There was an air of crazy unreality about the place, like a Mad Hatter's tea, with Rasputin doing strange things. He rambled aimlessly around the rooms, singing, dancing and talking, or he poured wine into tea glasses, sometimes not bothering to throw out stale tea inside, then he would bite some sugar and fruit and wash the concoction down in a gulp. The telephone constantly rang and he enjoyed talking on it, hinting that big deals were always pending between himself and some nameless ministers. On this day in 1915 he stopped such activities for a few minutes, long enough to greet his visitor from Moscow in friendly fashion. Then he launched a clumsy attempt at seduction, telling her that he might be able to honor her request, that he would use his influence in the government. However, there were conditions. When she demurred about her part of the proposed bargain he cried out, "If I desire you it's because God wants it that way and it is a sin to refuse. Without love I lose my power. You take away my strength. Give me a moment of love and my power will grow, and your affairs will go well." When he failed he turned violently boastful and sneered, "I have my women who love me and please me. Get out of here, get out!" he yelled. Then he turned to the telephone for solace. The next day he called and meekly apologized for his harsh words. But that was the end of the affair.

When he was sober he was clever and nimble-witted and able to take care of his interests, when he was drunk he was easy to deceive and cajole into doing almost anything. There were persons eager to help him stay drunk. He was one of those fortunate people who visited restaurants and cafes every night of his life and never had to pick up a check. With this debauching came another interest: politics, a dangerous pastime for him, one that brought a great deal of public attention. Normally, he preferred privacy, but a demon had seized him and his life was driven forward by the classic blind impulse that led to his ruin. People watched him and wondered what he was; at first they thought he was a Harlequin but then they were convinced he was Caliban, whose career was a triumph of moral and spiritual depravity.

He moved in a world he hardly understood. In the beginning he knew much less about politics and business than religion. Friends convinced him he would be guided by the special gifts that had marked him in his native village. He came to rely on the powerful intuition and dark energies of his mind to guide him in this new world. His supposed gift for knowing what people were thinking was largely illusory, but when it worked it was probably the result of a sensitive intelligence rather than any mystic prescience. After all, his personality and intelligence were late in developing and coming to consciousness, and he remained throughout his life a shrewd but uneducated man; therefore he could not discuss his thinking processes in rational and sophisticated language.

He tried to grope his way into a political outlook, but this political views were always hard to classify. Much of the confusion about his beliefs result from his secretiveness and from the fact that most of the categories of modern political discussion do not fit people like Rasputin. For instance, men such as Metropolitan Pitirim, a corrupt placeman despised almost universally in the church, was not a reactionary but an operator.

The social gospel element in his religious view carried over into his politics, but the Left ignored or dismissed him. Although he disliked landlords and bureaucrats, he thought the power of the state should return to what it had been before 1905. As a result the Right at first mistakenly thought he was one of them. Furthermore, he disapproved of the Duma. But his remarks did not give comfort to conservatives. "Are these the representatives of the people? No! They are great proprietors, aristocrats, rich people… There is not a single muzhik among them!"[21] We do not know what he thought of workers but he remarked many times that the autocracy should support the people, that is, peasants instead of landowners and presumably, workers instead of proprietors. He con-

---

21. Rasputin, *My Father*, p. 103.

sorted with speculators, trimmers and confidence men rather than businessmen or industrialists.

Many of the political figures near him did not know that on the subject of religious toleration he would have agreed with the Leftists in the Duma—if any of them had bothered to talk to him. Although he had friends among the leaders of the Black Hundreds, he could not accept many of the ideals of these organizations. He did not like their identification of church and state. His common sense saw that Russia was a multinational and multi-religious empire; many of the people in it felt alienated from official Orthodoxy and therefore were hostile to the national idea. It is probably too much to say that he was an active tolerationist; he feared the kind of trouble persecution brought on, and he had some vague realization that the empire did not need more tension but less because it was threatened by revolutionary outbursts. He feared where the mob actions of the Black Hundreds might lead. Although he was personally capable of loud and drunken belligerence he was a man of peace who had genuine dread of violence and its accoutrements. A noisy crowd, angry words in the Duma, or a harshly worded newspaper editorial might send him fleeing from the capital. A cautious peasant always fled danger rather than confront it.

Anti-semitism, a common bonding agent among many groups on the Right, did not interest him. He was inclined to think of the Jews as persecuted believers who had their peculiar ways, but he found nothing offensive in them. They worshipped God in their own fashion and therefore they should be left alone. At one time he opposed erecting a certain Jewish monument in Kiev. He probably feared that the city being a hotbed of ultra-monarchist political agitation, the presence of the monument might trigger rioting. In the capital some Rightists took alarm that he associated with sectarians and Jews and even tried to help some of them—for a price, of course. He liked Aaron Simanovich, a Jewish businessman who called himself Rasputin's secretary. Rasputin trusted his advice and placed many of his confidential affairs in his hands. In this entire question the final verdict was rendered, as far as anti-semites were concerned, by a Nazi author in the 1940's who called him "a tool of the Jews."[22]

The total effect of his activities and his ideas—as the public understood them—was to make him widely hated. He was a man with few friends, and he knew it. He was troubled but could not understand why

---

22. R. Kummer, *Rasputin, ein Werkzeug der Juden* (Nürnberg, 1939).

people reviled him. When this sea of hate began to swirl around him he fled into an endless orgy of drinking and pitiable attempts to play the great influencer. No one knew better than he how people lied about him, but he had helped make his own reputation and there was nothing he could do to correct impressions.

§

In order to understand Rasputin's political role in 1915-16, and the way in which he operated, it is important to keep in mind that he depended on a skein of relationships between the tsar, tsaritsa and himself.

Behind the belief that he controlled the state lurked several false assumptions. It was thought that he dominated the empress and emperor. But two barriers stood in his way. Although he was able to influence her at times he could never use her. Secondly, Nicholas refused to concede him any power. In the political sphere the will of the emperor was the most important factor. It was often muted, it is true, but only by the practical demands of politics over which he had no control. He accepted Rasputin merely as a kind of minor court chaplain who could offer soothing words to ease the worldly cares of a tsar's life.

His words written to Alix in August-September 1915, are the basis of the contention that he accepted her as virtual co-ruler. The meaning of these remarks demands more investigation than it has in the past received. At least the statements could not be taken at face value; they ought to be examined in the light of the emperor's reactions to her advice and meddling. When we do this a new picture emerges. We cannot know what prompted Nicholas to write these lines—perhaps a burst of affection or a moment of fear brought on by a sudden awareness of the new responsibility that he had taken on at the Stavka—but we should not be misled by them. In another, and more appropriate sense, they seem only to invite her to be a kind of administrative assistant who watched and reported events in the capital. But she did not envision herself in such a mechanical or passive role; she even spurned the notion of herself as peacemaker among the ministers. In her eyes they were palace servants, harmony among them being a matter to concern the emperor, who had only to issue a stern command that they be silent. She wanted to advise her husband and to participate in decision making. Her goal was not to be a mere helpmate and provider of comfort when he was distressed.

She had little faith in him and constantly deplored his faults and frailties; only a strong hand and firm will could guide the state in the crisis that followed the great retreats of 1915, and she felt that only she had such strength. Therefore, her advice and words had to be heard and taken to action. She was a Lady Macbeth who, as she piquantly put it, wanted to poke her nose into everything, to issue orders, to be obeyed. She yearned to exercise power, to see people tremble when commanded. Unlike her husband, who dreaded to use the might of autocracy, she was not afraid of power; it fascinated and intoxicated her.

She warned Nicholas to listen to her. As part of a campaign to intimidate him she invoked the name of Rasputin, hoping that her husband would recognize him as a Friend of God. Nicholas was aware of the elaborate theory she entertained about who Rasputin was supposed to be and what his presence meant; however he personally assigned Rasputin a lesser role.

It was her hope that Nicholas would accept her assistance, that he would accord her recognition as the immediate source of his best advice and that he would act upon it. She claimed the right to such powers because she could call on Rasputin, whose inspired words would reveal the will of God. Thus, through her, Nicholas could have access to wisdom that was not available to other mortals.

To all these entreaties he listened with patient, almost dumb willingness, and with a meek acceptance that for a time deceived his wife and continued to deceive observers. The noise of her clamorings caused many to overestimate the impact of her words. Nicholas bore these loud entreaties and aggressive prayers with the stolid patience that he bore other tribulations that were visited on him.

He calmly refused almost all she asked. In his household he was not the *roi fainéant* he was thought to be. He accepted the advice of Alix and her friend only when no other alternative was open to him. To grant many of her requests, Nicholas would have had to share power with her, and this he refused to do. He believed he was morally bound by his coronation oath not to share power with any other being; consequently, from time to time he reminded her that although he might hear her out the decisions ultimately had to be made only by him. He often showed little interest in her most cherished plans. He preferred to turn to her for the balm that soothed his spirit. Mostly he wanted to feel the warm fire of her passionate, obtuse self-confidence and obduracy. But the flame both attracted and repelled him; he drew close, only to flee.

Rasputin could get close to the power that these two struggled over, but he could manipulate it only in a minor way. He was, in fact, much disillusioned with the emperor. Once he warned Goremykin never to trust the tsar, a man who could promise anything, make a reassuring sign of the cross, but then fail to live up to his vow. When friends asked him to assist in a project to induce the emperor to appoint only Rightists to the upper chamber, the Council of Empire, Rasputin gloomily rejected the plan as hopeless. "What damned good is it? Right and Left are all the same to papa."[23] The tsar, in short, in the opinion of the *starets*, could not even be relied on to support people or policies that would help him.

The empress was the target of attempts by intermediaries trying to influence imperial policy, especially appointments, and it was thought that she could best be approached through Rasputin.

It is this connection with the empress that made Rasputin's position a complicated one. Its ambiguity was both his strength and his weakness. She was dominant and she usually bent him to her will, but he was a clever improviser and opportunist. He used the actions of the empress as testimonials of his power. It was easy for him to create the impression that he was the controlling partner, that he originated ideas and saw to it that they were carried out by her, who, he alleged, obediently put her power at his disposal. But he did not deceive himself. He knew perfectly well that his relationship with her was not accurately reflected by what she occasionally said and by what he was happy to repeat in public. Despite his position as Friend he was not free to dispute her.

Much of the misinformation and suspicion about his alleged power and his role at court derive from ignorance about the peasant cult of the Man of God. The people who wrote about him knew little of such things and made their guesses on the basis of ignorance. One of Russia's greatest historians and leading statesmen of the constitutional era, Paul Miliukov, rose to speak in the Duma in the spring of 1914 and in attacking Rasputin claimed that "The church is in the hands of the hierarchy, the hierarchy is the prisoner of the state, and the state is dominated by a common tramp."[24] He failed to draw a distinction between the word for tramp (*brodiaga*) and the word for the religious Wanderer (*strannik*). To him they seemed to be the same thing because he relied on appearances alone to make a distinction. The charge was unfair, but it is typical of the confusion arising about Rasputin because of ignorance of what he was.

---

23. Alexander Kerensky, *The Road to Tragedy* (London: Hutchinson & Co., 1935), p. 61.
24. GDSO, IV, 90.

As a result of this ignorance suspicion was piled on suspicion, and half-understood facts and misinterpreted facts were added.

From such ignorance a picture was created describing what he was and how his power functioned. He was a *starets*, they said, and therefore "owned" the soul of the empress. She was a slave to this cunning *muzhik*, and she gladly took his advice on all things. She in turn passed on the orders she received to her husband. In the *Admonitions* of the Monk Gregory of Zarub there was a warning that a person should never change his spiritual guide. Therefore, Alexandra was trapped in a permanent position of moral inferiority and subservience to Rasputin, and had to surrender her will to his wishes. This was the price she had to pay if she wanted to go on the road to perfection. The *starets*, if he were to be effective, had to be free to ask about her most intimate secrets; this gave him power over her and made her helpless. She could not resist any of his demands. She was a spiritual cripple. The popular guessing about the relationship went further: Rasputin said he could pierce her consciousness, see her secret sins, and uncover her most intimate weaknesses. She could hide nothing from him; he devastated her with his knowing, and promised that only he could prepare her for the freedom from guilt that these sins brought—she would just have to obey him.

§

Persons who entertained this grotesque tale and thought it was true did not know the empress. She did not accept Rasputin as a *starets* but as a Friend of God, something significantly different. But to the upper classes all forms of popular piety were a bore, mere different names for the same kind of superstition and ignorance that characterized the faith of the masses. To Alix, Rasputin was an extraordinary man, such as Phillippe had been, a vehicle of divine knowledge. She observed that his beliefs were like her own: he was indifferent to ritual, hierarchies, and the institutional aspects of church life. She accepted him as a seer, a healer of sorts, and a provider of wisdom. Eventually she made a leap in thought by considering him capable of advising her not only in personal matters but also in political ones. However, she took his so-called advice in a very special way. Although she believed he was an unconscious possessor of truth, she never thought that the words pouring from his mouth represented ultimate wisdom. She did not plan to gather up these sentiments of his and bring them to the tsar. If this was to be her role she would be nothing but a passive medium who could be replaced easily at any moment. On the contrary, she regarded herself as more indispensable than Rasputin. She proposed to extract knowledge from him, filter it through her mind, shape it, and then pass it on. Only she could interpret what he said: it was she who gave value to his words. This was her role.

Nicholas sometimes sought (and needed) the comforting words of Gregory; she did not. She lived well without them. Of course, when he died, she regretted his passing and even lamented a bit, but she was not devastated nor did she for a moment retire from life or suspend her furious activity. She was quickly taken up with a new cause of insisting that Nicholas severely punish the killers. All she wanted from Gregory was his advice on certain subjects. But without her to dress it up and carry it to the seat of power it was useless—words shouted on the wind.

She thought she was morally and spiritually superior to most other beings. Arrogance prevented her from listening to advice or criticism. Had Gregory presumed to suggest that he personally could instruct her she would have reacted violently against him. She was not interested in the advice of a peasant *starets*. That would have involved interference with her private life—something she would not tolerate. Over the years Gregory studied her and tried to offer her pious religious sentiments. He did not dare presume to make political remarks until the last year or two of his life.

Dealing with her was usually difficult for him. When she was certain of something, she was certain beyond an ability to reflect and to question. Her intense faith in her own convictions was something even Gregory feared to go against. He judiciously agreed with her frenzied importunings and in many situations assured her of the rightness of her opinions. She was not so smug as to think that unaided she could discern God's will. Without fully realizing what she was doing she forced Gregory to agree with her. She compelled him to assure her that God had approved her plans, then she was sure that her recommendations were the same as the will of God. First came her conviction, then Gregory's cautious endorsement. He could not do otherwise, giving back her own desires dressed up in the form of revelation which only a Friend of God, she thought, could perceive. The strength of her personality caused Gregory eventually to admit that what she almost hysterically wished to be true really was true. If God had in this indirect way sent advice she did not want to hear she would probably have been hostile and suspicious toward Gregory.

She looked into his mind and saw her own faithfully reflected desires; therefore she thought she was viewing the face of wisdom. He discovered and revealed to her few new facts, in most cases he was merely used to confirm what she had intuited. In a few instances when he clamored on behalf of one of his own favorite projects, she showed scant interest or cold indifference. She was never completely convinced about his fear that the food crisis in the cities could lead to revolution, and therefore she was not too helpful in this matter, For the most part she pushed her own proposals. When they met with resistance from the tsar she armed herself with blessings extracted from Rasputin and then cast aside the slender shreds of doubt and offered her husband the opportunity to choose the right path, the one she had marked out. Success deepened her feelings of righteousness, failure disturbed them not at all.

§

Rasputin was the only member of the strange threesome who understood approximately what was happening, and he was also aware of the nature of his power. He knew that much of the popular reputation he had acquired depended on public ignorance, which led to incorrect surmises about his position. In general it can be said that the role he played in tsarist politics was partly an illusion. The beholder, his senses stimulated by propaganda, was nevertheless convinced that he was seeing reality. But Rasputin was not deceived. Of the ministers he asked only that they concede him the shadow of power, which he might use for his petty and sometimes sordid projects. Goremykin and Sabler deferred to him in this respect. However, they aroused the wrath of the Duma, won reputations as hirelings of the *starets*, and were cursed by the public. But they pleased the empress and removed her annoying, noisy meddling from their offices.

Help from the ministers did not end his troubles and dangers. He had no real power, but he had to act as if he did. His course of action was at all times dependent on what was happening in the mind of the tsar and in the highest circles of government, yet his sources of information were not always good. He had to rely on conversations with the light-headed Vyrubova, who picked up scraps of information at the dinner table at Tsarskoe Selo, or who passed on word from the empress, who was not by any means privy to all of the things that her husband was thinking. Rasputin's problems were many; at any moment he could have destroyed his reputation with a wrong move based on misinformation.

The tenuous nature of his power is shown in the cautious way he permitted himself to become involved in the appointments. The reliability of his information was not superior to that which many hangers-on at court had. While he considered the qualifications of the office seeker, he needed to gather facts about him, confirm the impending dismissal and appointment, and calculate the movement of the emperor's thought. A close look at the ministerial appointments does not support the idea that Rasputin was the real power behind the throne. In these cases he failed to reveal a completely clear understanding of what was happening, although he showed a remarkable ability to protect his own interests in difficult circumstances. He had not the firm grasp of affairs that the successful intriguer might be expected to have. He was unable to bring pressure to bear at the right time and in the right place—that is, on the emperor.

After having consulted his own interests and after having surveyed with care the ground before him he very gingerly entered the turmoil involved in an appointment. He always had his own interests, and they were never completely paralleled by those of the candidate. He had to be careful, but not because he was concerned with the nuances of a candidate's political outlook. He had his own principal and secondary reasons for taking part in appointments; most of these reasons were matters of expediency. In many respects he agreed with the Duma, and with the tsar, that the candidates were a poor lot, and most of the creatures he supported had only his limited sponsorship because there was no other person available for the post. We never see him going all out in his support for any man.

He did not have to concern himself with a candidate's political beliefs. His test of loyalty was a simple one: Did he love the tsar? The very fact that the man or his agent approached Rasputin was a good indication this test was already passed, since the *starets'* reputation was a beacon of light which attracted only the supporters of autocracy or those who were corrupt enough to support anything provided they got power. Politics were not his concern. It was his job to find someone (after 1915, almost anyone) who would serve the regime. He seems to have been cynical about the possibilities of finding men who could sincerely pass as devotees of autocracy and who would concern themselves with his own small projects.

Rasputin never understood the game of politics nor did he seek to play it. Near the end of his life, under the tutelage of Manasevich-Manuilov and others, he was learning some of the political facts of life and coming to realize that his own goals could be realized through participation in politics. But it was too late; his viewpoint remained that of the typical peasant *starets*. He retained a primitive notion that office meant power, and the wielder could help people by granting their wishes. He wanted largesse or access to sources of influence that would help him to distribute some of the small favors that were his principal stock in trade.

The charge that he was the power behind the throne must stand or fall on the basis of evidence about the role he played in ministerial and other appointments. Basing an estimate on the letters of the empress and the testimony given by the former members of the imperial government, we can say that Rasputin was interested in about forty nominations. Many of these were minor posts such as the governor-generalship of St. Petersburg. In some of these cases he had vital interests at stake, but in others he was merely mentioning names in order to please the empress.

If it could be proved that he was able to bring about nominations and appointments of ministers, then it would be logical to assume that he could appoint other people to other posts. But could he hire and fire ministers? A number of these nominations are not sufficiently documented to permit clear judgments to be made. If we can choose a few cases for which there is a fair amount of documentation, we can see where Rasputin fit into the picture and what he was trying to do. None of these cases show him at a low point of activity; in fact, they appear to have been deliberately scrutinized by the investigating commission of the Provisional Government because of all the cases they tend to show Rasputin in his most active role. However, they fail to confirm the charge that the government was in his hands, that he was the single most powerful person on the political scene, or that he actually ordered and used the tsar and tsaritsa for his own purposes.

In none of the cases examined did he make a minister, nor was his approval the principal reason that an appointment was made. A study of the episodes leaves one with the conviction that a revision of former viewpoints is needed. It is by no means necessary to discard all of what has been said about him in the past. But it is now apparent that Rasputin's influence was much more restricted than has been supposed. His story is about the limitation of his role and frustration of his hopes.

There is still ample room for a discussion of the extent of his power. Such matters can never be described with perfect precision. But it seems evident that Alexandra was not in control of her husband's political judgment; and it is even more clear that Rasputin did not control the empress. Still, contemporaries insisted he was the master of Russia and they finally misled themselves by their talk and speculation.

VIII

The Making and Unmaking of Ministers

The imperial government went about choosing ministers in a way endorsed by tradition. Theoretically, the tsar was not bound by legal restrictions in his choice; he could pick anyone, and he was not held accountable. In practice each nomination was preceded by a period of discussion among the important members of the bureaucracy, and the Senate and Council of State might informally consider the claims of the leading candidates.

In theory a candidate's program or views were thought not to be important, for all ministers were regarded as servants of the autocratic tsar, who framed policies and then ordered his assistants to find the best ways to carry them out. But even under the autocracy of Alexander III men were known to represent beliefs and were picked because of what they thought as much as for what they were able to do.

However, after 1905 it became increasingly difficult to understand the emperor's reasoning and to see whom he was consulting and how he was arriving at a selection of men for high offices. Onlookers were confused and made all kinds of guesses about his motives. For a long time answers eluded them. When the World War and its attendant crises sped up the process of hiring and firing ministers, the problem became more acute and the attempts at explanation sometimes were far-fetched and fanciful. In such an atmosphere it seemed probable that Rasputin might be the power behind the "ministerial leapfrog," as it was called. In some cases the appointees themselves did not understand why they had been named. Many of these resisted and begged to be let off, but the emperor was relentless. To an interviewer, A.A. Rittich once expressed complete mystification when asked why and how he had been chosen Minister of Agriculture. Asked why he had been fired, he only guessed that Rasputin had something to do with it. Several factors made recruiting difficult in the second half of Nicholas's reign. The requirement for ministers to appear before the Duma was a new and extraordinary demand placed on the corps of men who had been trained under the old system of statesmanship where such an ordeal was undreamed of. Many could not (and would not) care to go through this frightening experience. The Russian constitution (like the German) had a cabinet responsible to the sovereign, not to a parliament. The Duma, however, had weapons it could use in a confrontation with a member of the Council of Ministers who refused to be cooperative. He could be harassed with interpellations on the floor, he could be upset by fretful questions before the commission concerned with affairs of his department, or he might find his budget appropriations held up. The emperor did not realize how much moral courage was required of a man who would have to face hostile deputies. The minister had to carry out orders without angering the Duma. But if he retreated from principle or tried to sound agreeable he risked losing the confidence of his master. Few men felt they had the ability to live in such a crossfire.

Even the most loyal members of the government knew that the revolution of 1905 might renew itself at some later date and then the entire question of the fate of tsarism might be raised again through revolution or *coup de état*. A new constitutional regime might be the result, and career conscious administrators knew that those who had associated with the old government might not be welcomed into the new one, so some were hesitant about serving the tsar lest their future careers be jeopardized. Many were standing off, waiting for the dust to settle so they could decide if it were prudent to serve. But before the World War such caution was found only among a minority of statesmen. At the time everyone, including the tsar, was aware of the possibility of revolution or considerable modification in the government. As years passed, especially during the war, the number of people holding this view grew larger, therefore the ranks of those willing to defer office holding grew larger. All this was not disloyalty to the regime; it was foresight.

During the last months of its life tsarism lost supporters for other reasons. The torrent of abusive gossip and criticism by the Duma and various public men had generally so discredited the regime that many persons out of conviction did not care to serve lest their personal reputations be blackened. The campaign of vituperation ran so strong that the man who accepted a post found that his friends regarded him as a traitor. This kind of pressure had the effect of removing most eligible people from the pool of candidates for office. Under such conditions tsarism could recruit only the timeservers and the amoral.

These conditions placed a cruel problem before Nicholas. He had to behave as if candidates were in abundant supply. Formerly, he could fire a man with the knowledge that some kind of successor could quickly be found, and there was no great hurt to the government if one were not discovered at once. Under the new conditions Nicholas had to maintain a pretense that his government still had wide support. This meant that a post could not be left vacant a long time for fear of the results of gossip. Before a man could be dismissed, his successor had to be found. Officials had to be kept in office after they began to ask for dismissal or after their services were not wanted. Once it was decided to fire any important official, a frantic search got under way to find the new man. The old theory was used to explain what was going on: the tsar was merely looking for a capable replacement and there were plenty of aspirants willing to serve. But actually there were few who wanted to be known as willing tools of the autocracy, which every day was losing friends and becoming more and more politically disreputable.

In this dilemma the tsar began to act desperately in order to find men. There was little time for careful consideration of candidates. Rasputin was close at hand, and there were rumors that he was suggesting or even ordering the appointments. Rasputin took advantage of people's credulity and claimed that he was indeed providing the names of the men who were being given high offices. This boast was accepted at face value; no one thought to ask what was the precise role of Rasputin. No one saw that his role was not only vague but also was a minor one.

§

In the summer of 1911 Alexei Khvostov, the governor of Nizhnii Novgorod, was busily supervising the great fair that annually took place in that city.[1] A young, ambitious aspirant for high public office, he had already served as governor of Vologda and had several times been received by the tsar at Tsarskoe Selo. In fact, his most recent visit had been only six weeks earlier. There were rumors heard in high administrative circles that the Chairman of the Council of Ministers and several of his fellow ministers were soon going to be dismissed and that the emperor was already looking for men who could fill the posts. Khvostov was lost in pleasant reverie as he considered his own qualifications for a cabinet position. For one thing, the emperor had been very cordial during their conversations, and at one time had hinted that he had his eye on the governor and was considering him for a post of importance in the future. Not that Khvostov was dissatisfied with his present position; the governorship of the ancient town was regarded as symbolically important, for the place was rich with historical memories. When this governorship was given to a young man it was a sign that he was a statesman with a future. Khvostov sensed he had done everything to insure his future in this way. In the capital he had visited and cultivated whatever influential persons he had been able to meet and exhibited himself as a devoted monarchist and prominent member of the Black Hundreds and other patriotic organizations.

His musings were suddenly broken off by the entrance of a secretary who announced that there was a visitor, a peasant. The governor prided himself on the fact that his door was always open to all men rich or poor who wanted to see him. He consented to talk to the visitor although he was busy and impatient to get back to his work. The peasant appeared in the doorway, was motioned to a chair, and sat down. In his attitude he was deferential and yet business-like. He was intent on his business but nervous, and from time to time he shifted his feet which were clad in highly polished black boots.

Through the conversation the visitor never cast his eyes shyly down in the way of most peasants when in the presence of an official; instead, he fixed his gaze on the governor and announced that he had come "to observe his soul."[2] Khvostov realized that he was in the presence of some kind of holy man whose strange language was freely sprinkled with allusions to the fact that he moved in important circles in the capital, and he intimated he was on an official errand. Although he was restrained there was an insouciance about him that was disturbing. Gradually the governor understood that Gregory Novy, as he called himself, was Rasputin, a shadowy character whose name was mentioned in the capital. After they

---

1. *PTR*, Khvostov, I, 1-53.
2. *PTR*, Khvostov, I, 3.

had talked for a while Khvostov was friendly but firm as he began to usher the visitor toward the door. Novy asked to meet the governor's family but was put off with a protest that the governor had a busy schedule. Then the visitor hinted that he had been sent to look at Khvostov to decide if he were suitable as a candidate for the post of Minister of Interior. During the remaining few minutes of the visit, Khvostov treated the incident as a harmless but annoying joke. This was the first meeting of Rasputin and Khvostov.

Thus ended Rasputin's initial adventure in politics. For several months he had been friendly with G.P. Sazonov, former liberal and currently an editor of a conservative journal. Sazonov was the first person to try to use Rasputin for political purposes. He had introduced to members of conservative groups in Kiev. In the summer of 1911 the *starets* was much in need of friends, so he accepted all proffered hands. He and the journalist huddled over a number of business and political schemes and both went to Nizhnii Novgorod to get the help of Khvostov.[3]

Sazonov took at face value the gossip of the salons and the newspapers which rumored that the *starets* had considerable influence at court and his word could assist in the making of appointments. The almost indiscernible figure of Witte, like a star of some small magnitude, lurked in the background of the plot that sent Rasputin to Nizhnii Novgorod, for Sazonov sought to use the pure and undefiled conservatives of Khvostov to make palpable to the emperor an arrangement which would permit Witte, a radical in the eyes of the emperor, to find his way back to power. Khvostov was to be proposed as Minister of Interior and Witte as Chairman of the Council of Ministers.

These impractical maneuvers called for considerable knowledge of opinion at court, and none of the principals had this knowledge. Witte was not aware of the extent of the emperor's hatred for him, and Sazonov and Rasputin were equally unaware of the true situation at Tsarskoe Selo, although Rasputin could probably have made a better estimate than anyone of the state of affairs there. He had been used as a pawn, but since he was totally unable to influence the course of events it was many years before he again appeared in politics.

For the moment he was swept up in the fight over the appointment of Minister of Interior. Nicholas II wanted Khvostov considered, but he was more interested in Nicholas Maklakov, the governor of Chernigov. Kokovtsev, the premier, was pushing the cause of A.A. Makarov who had

---

3. Witte, III, 564-68.

served as the Minister of Justice and Interior.⁴ At the suggestion of the empress, Rasputin was dispatched to Nizhnii Novgorod. Whatever the emperor thought of this junket, he would never have accepted a man whose name appeared on the same slate with Witte; therefore, the plan of Kokovtsev won out and Makarov was appointed. Rasputin, somewhat bewildered by politics and hoping to make a friend, pronounced the new minister a good man although he had never met him.

During the next six years Khvostov worked at his trade. In 1912 he cooperated with Minister of Justice Shcheglovitov to make sure that the electoral books of Nizhnii Novgorod were carefully edited to assure the defeat of the local Constitutional Democrats.⁵ At the same time he arranged to have himself elected to a seat in the Duma, and he soon became one of its staunchest Rightists. In the spring of 1915 his faith in autocracy wavered and his voice was heard among persons protesting the alleged German influence in Russia—an oblique way of striking at the dynasty. Fortunately for him his attacks progressed no further than harsh words aimed at a number of banks. Then once more Rasputin and opportunity knocked on his door. This time Khvostov was ready.⁶

The great military defeats of the spring and summer of 1915 brought on a massive political crisis. Nicholas II responded by dismissing some of the most hated ministers, Shcheglovitov and Sabler among them, and appointing several men acceptable to the Duma. As the summer drew to a close it became obvious that the German armies were not going to accomplish their goal of knocking Russia out of the war; consequently, the emperor resisted the demands of the opposition for a ministry of confidence and official acceptance of the Progressive Bloc of deputies as a national party. The crisis caused by the clash over these proposals was soon followed by another crisis: Nicholas announced that he was taking over command of the armies. For a number of different reasons there was almost universal resistance to the idea. Ten of the ministers led by S.D. Sazonov, Minister of Foreign Affairs, signed or approved a collective letter in which they demanded the emperor's plan be abandoned. The letter sounded like a threat of mass resignation of a parliamentary cabinet. Nicholas had to get rid of the rebels, but he had to do it piece-meal lest he add to the impression that these were parliamentary ministers who could resign because governmental policy differed from their own private views.

---

4. Kokovtsev, p. 277.
5. "Iz arkhiva Shcheglovitova," *KA*, XV (1926), 1070-8.
6. *PTR*, Beletsky, III, 255-301, 327-432; Andronnikov, I, 361-402, II, 9-23.

It was apparent that there would soon be vacancies in the council. There were people who regarded this as an opportunity that ought not to be wasted. One of these was Prince Andronnikov, one of the most loathsome persons who existed on the periphery of the government. A noted homosexual, he had dabbled in journalism and politics and had from time to time been employed as an obscure office holder in the Holy Synod and the Ministry of Interior, where he had served as a receiver of bribes and as an informer. He was the perfect example of the confidence man who was ready to profit from any kind of shady deal. For some time he had been posing as a man of influence. His technique was to gain the trust of a servant close to the tsar. This person was supposed to glance at the papers piled on the tsar's desk. The information as to what business was pending was given to the prince who then called on the individual whose affair was due in a short time for imperial consideration. Andronnikov informed him that action on the matters he had before the tsar was about to take place. Another tactic of the prince was to learn about an appointment a few hours before it was made public. Then, carrying an arm full of flowers or a box of bonbons as a gift, he called on the successful candidate and congratulated him on his new position. This would be the first time the appointee learned of his success. Both these tactics convinced people that Andronnikov had played some sort of role in expediting business or in making appointments. The prince usually was willing to accept some gift or favors for his supposed assistance.[7]

Andronnikov was close enough to the government to know that it always experienced difficulty in finding men to serve in the council and he realized that the threatening situation in August 1915 would present a grave problem to the regime because there were many appointments that had to be made soon. The field from which the choices could be made was extremely limited. Only a relatively small group of high ranking officials was known to the tsar and among these there was a general reluctance to take a position in the council. The problem then, as far as the prince understood it, was to find some way of bringing the isolated tsar into contact with a man who made himself qualified for a post merely by being willing to serve in it. Andronnikov knew of Alexei Khvostov and his long-standing ambition. He decided this was the man he would make into a minister, and perhaps with good fortune, into a Chairman of the Council of Ministers.

The prince believed that he understood the process by which nominations were made. Nicholas usually cast about among the circle of his

---

7. Miller, pp. 102-4. Mosolov, p. 165.

limited contacts in an unsuccessful attempt to discover a candidate. Then from the imperial household began to flow suggestions and names of potential candidates. The prince knew that this information was sometimes brought to the empress by her friend Vyrubova and by Rasputin, who was known to Andronnikov since 1914.[8] But he thought that Vyrubova was the more important of the two, so he concentrated his attention on her by sending frequent notes praising Khvostov and suggesting that he was available for a post in the Council.

He did not expect Rasputin's role to end once the appointment was made. He knew that one of the things that could make a minister ineffectual was a lack of information about what was happening at court. Andronnikov saw that it might be possible to secure information of this kind from Vyrubova and from Rasputin, so he prepared to use Rasputin as a source for all kinds of court gossip. In return he thought that Rasputin would be made happy by small sums of money and gifts. He believed also that Rasputin could continue to bring kind words about Khvostov to the empress.

The prince's assumption that a minister could be made by the kind of intrigue he was carrying on was partly correct, but he had not learned the proper combination of tactics to bring about the fulfillment of his wish. Long after the empress was completely convinced that Khvostov was a worthy man, the conspirators directed their efforts at her. They did not know that the emperor often made appointments which she did not like and that he could refuse to make those that she begged for. Numerous letters to Vyrubova, most of which were shown to Alexandra, failed to secure the appointment. As the weeks went by the conspirators asked themselves what was wrong. More and more their attention began to focus on Rasputin.

Much of the press, reflecting on a wide range of opinions, was hinting that all the ills of Russia could be attributed to Rasputin. He was undermining the liberal ministers, he was behind the effort of the emperor to take command, he had advised the emperor not to give in and to reject the program of the Progressive Bloc. Khvostov, who heard these rumors first-hand in the Duma, and who heard Rasputin's boasts, was convinced the *starets* stood behind the throne. Therefore, he came to think that Rasputin was the key figure who had to be won over by promises of rewards. Khvostov also knew that Alexandra was thundering against Prince N.B. Shcherbatov, the Minister of Justice.[9] The prince was a man of moderate

---

8. Spiridovitch, *Raspoutine*, p. 252.
9. A to N, June 24, 1915.

political views but he hated Rasputin. As Minister of Interior, Shcherbatov controlled the police, and if he cared to he could have removed the guard from Rasputin for a few hours. This would surely mean the death of the *starets*. Khvostov was convinced that Rasputin was very important to the empress, so he tried to sell himself on the basis of his vow to protect Rasputin.

To bolster his cause, Khvostov brought in Senator S.P. Beletsky, who had until this time been rather inactive as a collaborator in the plot. Andronnikov, who knew Beletsky well, recommended him as Assistant Minister of Interior, giving him control over the police and all security measures designed to protect Rasputin. If the safety of the *starets* was something that would sell the appointment to the empress then Beletsky would be valuable because he could be presented to her as the man who would spend his every waking moment protecting Rasputin. Andronnikov believed that there were other reasons why Beletsky was important: he would have control over dispensation of police funds which the accounting offices of the Ministry of Interior could not completely oversee.

The two officials were presented to the empress as a package appointment which could solve many of the problems of the country. Before their plans were discussed both men were portrayed as being above reproach in their devotion to autocracy. Both were said to know Rasputin and to revere him.[10] Khvostov was even reputed to have defended him in the Duma. This was not quite true; he had only refrained from signing an interpellation that might have involved the introduction of Rasputin's name into debates.

Although Rasputin was part of this cabal, he observed its maneuvering with a certain degree of detachment. The three prime conspirators had different goals in their quest for office. Andronnikov saw that a successful plot would give him a chance for more profitable transactions than his previous engagements. Beletsky was interested in bureaucratic advancement and honors, and Khvostov wanted high political office in which he could wield great power.

Rasputin's aims were similar to those of Andronnikov's, although personal wealth was not a consideration for him. He was well aware that he could not associate intimately with any clearly identifiable political group; his high rating with the empress rested on his independence. As an independent operator he sought goals which had little to do with state

---

10. Khvostov made an attempt to win the support of Goremykin by appealing to his mild (and private) disapproval of the *starets*. He said he was deliberately trying to discredit Rasputin by getting him drunk in public places. Rodzianko, p. 158.

power. He wanted friendship with persons in high places only because they might help him to do the things he believed most worthy of his calling. He was, after all, a Man of God, and politics was a sordid thing to him. He never became enamored with Khvostov; he accepted him only because he was not "rotten with politics."[11] For himself he wanted to cultivate a reputation as the man who could get things done, who could take on the petitions of the most insignificant people to the tsar. By the summer of 1915 his apartment was the meeting place of all sorts of persons who came to seek help. Most were honest, some were not. Generals and aristocrats sat down in the waiting room next to beggars and unwashed peasants. Most of his visitors were humble folk who did not know how the government worked, so they came to him with their small problems, hoping he could help them. Rasputin was not skilled in uncovering the motives of his visitors, and he sometimes became involved in crooked schemes.

In the presence of all he jotted down on a slip of paper the name of the person and his trouble, and then he sent him with a written request for help to an official. Sometimes the paper was sent directly to the official. Every day these crudely written notes appeared in the administrative offices of the government. Most of them were ignored, although each official who ignored them was certain that all of the other recipients were obeying these pieces of effrontery. The truth was that Rasputin had a few friends in the government who assisted him in a small way, but almost no one of importance helped him, and he knew that he could not pass these requests for personal favors to the empress. To her he sent other and probably equally innocent petitions about topics he regarded as important, such as the critical shortage of supplies in the capital or the large horde of unhappy refugees squatting in the suburbs.

Rasputin was unable to use his friendship with Goremykin. Although they had met only a few times and the old man had expressed admiration for the *starets*, there is reason to doubt his sincerity. But the statement cost him nothing because he and Rasputin understood that it did not permit Rasputin to interfere in the government or even to ask favors of the Chairman of the Council. On the other hand this public confession of Goremykin's alleged liking of Rasputin meant that Goremykin could gratify the empress. With the chairman Rasputin had little influence; therefore, if he wanted to be a man of importance he had to strive to find in the council a man who would work with him.

---

11. Rasputin, *My Father*, p. 102.

Rasputin never gave himself completely over to this plot between the other three. Aside from a desire for independence, he had retained a suspicion of the people with whom he was dealing. From Vyrubova's gossip he was able to perceive what was happening. The plotters had won over the empress without much difficulty, but the emperor remained unconvinced. In desperation they were turning to Rasputin for help. He invited their requests for assistance by boasting of his influence. Once the trio was convinced that he had to be an active partner they were willing to pay a high price for his aid. In an interview, Rasputin scolded them, pointing out that he was aware they had sought to consummate their drive for power while he was away in Siberia during August of 1915, and that they had generally treated him as unimportant. He recalled Khvostov's snub in 1911. Khvostov was properly apologetic and said that he had been young and inexperienced at the time. He also countered by saying that if Rasputin wanted protection for his life he had to have an absolutely trustworthy police escort, something that only Beletsky was ready to guarantee. This placed the discussion in the proper perspective, in which both parties acknowledged that they had things to gain if their plot was successful.

From this point the conversation moved on to money and information, the things which interested Andronnikov and Rasputin. They agreed on an elaborate treaty in which the participants carefully outlined what each would contribute and to what each was entitled. Khvostov and Beletsky were to receive from Rasputin all kinds of court gossip that would be useful to them. Rasputin in return would receive 1500 rubles per month in pay, and the ministers would seek to do for him whatever favors were possible. Both information and the money were to pass through the hands of Andronnikov who would serve as intermediary between the parties. However, the foursome arranged to meet fairly frequently in secret conclave. Finally, Rasputin agreed to make a special effort to keep Khvostov informed as to the activities and opinions of the emperor.

The *starets* had promised much more than he was capable of delivering. He had almost no contact with the emperor. He had to rely exclusively on Vyrubova to tell him what the emperor was saying, and she was a very poor source of information because she saw Nicholas only during meals when the entire family sat down together and when all discussions were generally light table talk. In addition, the letters of the empress reveal that Rasputin seldom tried to submit at court any of the requests for favors that the other plotters asked for. The later history of this scheme, which ended in a political scandal, indicates that the conspirators did well to be suspicious of each other's intentions.

§

The emperor had at this time decided to take over command of the army. For a long time he yearned to lead his troops as his grandfather had done in the war with the Turks. At the start of the war against Japan, Nicholas said that his place was at the head of the armies in Asia. When the defeats began in the spring of 1915 his mind turned back to this plan. But the mere mentioning of it caused advisers to shrink in horror. Alix hesitated for a while—she was aware of some of the risks—but when she thought of the danger of permitting the ambitious Grand Duke Nicholas Nikolaevich to remain in control of the armies, she began to back the proposal strongly. The emperor was stubbornly determined to go ahead, but he dreaded the almost universal outcry that would greet his announcement. Alix stepped in to her usual role to lend him moral courage.

She felt this was a moment of great danger. Between June and August of 1915, she believed, the Duma had failed in its efforts to gain a strong constitutional voice in the government, and she feared that its next effort would involve an attempt to use the army to drive Nicholas off the throne. If, however, he took command of the armies the plot would be thwarted. But Rasputin was a bother to her because he warned against the emperor's plan just as he had warned against the foolhardy imperial jaunt into Galicia in the spring of 1915, a trip that had resulted in great embarrassment to the regime when the counterattacks of the Central Powers drove back the Russian armies. She worked with tremendous energy to encourage Nicholas and induce Rasputin to give his support. When Rasputin saw she was almost hysterically committed to the scheme he gave in and then pretended to support it with conviction. All this had cost her much effort and by the beginning of September she was weary.

In came a new threat and once again she was busy, fighting relentlessly in the cause of autocracy. She cried that the ministers who had signed the collective letter opposing the tsar's taking command had to go. For the moment she fastened her attention on Shcherbatov because she feared he might bring about the death of Rasputin. Andronnikov, through Vyrubova, had offered the services of Khvostov. The aroused empress was eager to see Nicholas begin the dismissals of the protesting ministers. While the emperor was home she talked to him repeatedly about Khvostov, and by the time he departed for the Stavka in late August she had convinced herself that Khvostov was the man to replace Shcherbatov. Only her anger and desperation forced her to accept Khvostov so early in the game when she knew little about him and could offer her reluctant husband few reasons why he should be named.

At the moment Khvostov had many things against him. For one thing his uncle, Alexander Khvostov, freely spread the word at the Stavka and in St. Petersburg that he was not honest.[12] Goremykin did not like him nor want him in the council; his loyalty to the regime was suspect since he had participated in the Duma debates about the alleged German influence in government. Alexandra said one word on his behalf: he liked Rasputin. This to her was an important test of any man's basic goodness, but even she understood that it was not enough to win an appointment. He needed the approval of Nicholas. Still, she felt he was the best man available, and she was certain she had convinced her husband of this. Consequently, when Nicholas returned to Stavka she requested him to wire her an agreed upon code word to let her know when he had ordered the appointment.

There was no other likely candidate for the post, but the empress had to wait for six weeks before Nicholas finally gave in. Once she almost gave up hope and indicated some willingness to consider Alexis Neidhardt, a member of the Council of Empire who had conducted an investigation of German influence in Russian business circles. Neidhardt indicated he believed in working with the Duma rather than trying to fight it, so his cause collapsed. He had never sought the post actively, anyway.

Nicholas refused to give in to her furious pleas. He waited and demurred as long as possible.[13] Several times he said that he had forgotten to discuss the matter of the appointment with Goremykin. Twice he put her off by saying that he would consider the case when he returned to Tsarskoe Selo, but each time he failed to act. To these obvious stalling tactics she answered petulantly, and when she refused to take any excuses Nicholas merely ignored her demands. To this she responded angrily asking if he was receiving all her letters.

She could not see that the emperor was only hoping that someone more suitable than Khvostov might appear; she thought that he was wait-

---

12. *PTR*, A.A. Khvostov, V, 467.
13. N to A, August 31, September 18, 1915.

ing because he was not completely sure of the man's fitness. Indeed, he was not. His opinions of Khvostov swung back and forth as he sought to make up his mind. Vyrubova had met Khvostov in company with Andronnkov and she warmed to him at once. But the emperor probably preferred the opinions of Goremykin who after an investigation decided against Khvostov.[14] From late August Alix gradually built up a case in which she thought she was demonstrating to her husband Khvostov was a good man. She noted that he had approved Nicholas's taking command—a fact, incidentally, that could not be proved. She said that he explained his involvement in the issue of the German influence as a ruse to forestall the Left in the Duma before it could take over the debates; he had a Russian name; he knew people and had a wide circle of contacts; he could control the Duma and the press. She knew that the emperor would probably be concerned with how the young Khvostov would cooperate with his uncle and Goremykin; both disliked him. The candidate got word to her that he would defer to the two old men and there would be no friction in the council because of these personal relationships. She reminded Nicholas of the danger in having Shcherbatov in office and of the pressing need to fire other unreliable ministers.

On September 17, Khvostov saw the empress in person. He had prepared himself for this crucial interview by picking the brain of Rasputin for several weeks and encouraging him to discuss the empress and her fears. As a result, when Khvostov talked to her he sought to assuage her fears, to discuss all of the things that concerned her, to show he was alert to them and had a plan for dealing with them. Khvostov's performance impressed Alexandra and her letters of that day reveal that he left her contented. There was a feeling in the lines that she at least had assisted her husband in finding the strong man they had dreamed of for years. Her praises rang with superlatives. Khvostov not only promised to carry out specific matters of policy but also divulged to her his beliefs about Russia and its future. He told her everything she wanted to hear and what she heard was a rare instance of a person who seemed to really believe in autocracy. Political figures had talked about their devotion to the monarch, but had avoided saying if they thought autocracy had a future. The consensus was that it might live on only if it were radically altered. But not Khvostov. The empress admitted that Russia was in crisis, and so did Khvostov. But he stated that there was nothing fundamentally wrong with the system of government; it required men of courage to direct it, and it

---

14. A to N, August 29, 1915.

needed planning and wise administration rather than change. No one had talked of long-range planning for politics because few thought that autocracy after the war would participate in the final solution of many of Russia's problems. Khvostov, however, discussed post-war problems, stating that it was important to get to work at once on ways of demobilization that would not create mobs of unemployed soldiers feeding the political maelstrom. The empress mistook the mentioning of these problems for an intelligent analysis of them. No one, not even her husband, had talked of such things because of the fear that tsarism as it existed in 1915 might never live to deal with them. The effect of the interview on the empress was overpowering. Nicholas said nothing.

It is obvious from the letters of the empress that Rasputin was playing a double game. He was pretending to cooperate with the plotters far more than he actually was. At court he was non-committal on the subject of the projects of Beletsky and Khvostov. He was interested only in seeing that Alexandra gave Khvostov frequent reminders that he and Beletsky had to make certain that the life of Rasputin was well protected.[15] The empress found it difficult to get from Rasputin any kind of enthusiastic endorsement of the pair. She said that Rasputin approved of Khvostov, but then admitted that she deduced this approval from a vaguely worded telegram he had sent from the Caucasus. Khvostov, in fact, became suspicious that Rasputin was not giving him all of the support he had promised, so he later rushed off to Tiumen to see Rasputin. He hoped to convince the *starets* that he was making a mistake; he also hoped that this trip would convince the empress that he was close to Rasputin.

Nicholas held out until the 26th of December when he finally named Khvostov Minister of the Interior.

It did not take long for the conspiracy to break up into a discordant charivari. The debates in the Council on the emperor's taking command and the signing of a collective letter of protest by the ministers signalled that there would be many replacements in the council. But the Khvostov-Beletsky-Andronnikov axis was not successful in getting its aspirants into the positions. They could not even prevent the appointment of an enemy, A.F. Trepov, in the post of Minister of Ways of Communications, and when they attacked the clever and capable Bark, Minister of Finance, he drove them away in fright by issuing a curt threat. The promise of power that had once seemed so great had shrivelled into nothing.

As a result of these failures the inevitable feuding and blood-letting began. First the trio examined Rasputin's function. They had been meeting him secretly and receiving some information from him, but it was useless. They suspected that he would not—or could not—keep his part of the bargain. He transmitted to Alexandra almost none of their requests for favors. The items she forwarded to her husband during the autumn of 1915 sometimes included references to Khvostov but these were mostly things Rasputin had a personal interest in, aside from the needs or desires of the minister. Rasputin was now deeply engaged in ecclesiastical politics, doing his part to remove Samarin from the Holy Synod, pushing the irregular canonization of St. John of Tobolsk, and favoring the candidacy of the corrupt Pitirim for Metropolitan of St. Petersburg.[16]

The ministers decided first that Andronnikov was sabotaging their efforts, and as a result their attempts to exert influence through Rasputin were failing. Andronnikov did not really have the power to do this although he was plotting along these lines. He was moving to make himself the principal agent of the *starets* so that he could control all business transactions between Rasputin and the ministers. This scheme of Andronnikov's caused the cabal to break apart. Rasputin agreed, on the demand of Khvostov and Beletsky, to ignore the prince, who was subsequently exiled to Riazan after the ministers ruined him by confiding to Vyrubova some of the scandalous truth about him.

Once in office Khvostov began to examine his position. The visits before him offered a number of attractive opportunities he had not seen before. First, Goremykin was trembling at the thought of having to face another session of the Duma. Khvostov believed he could offer himself for the post of Chairman of the Council. This move would bring him under the close scrutiny of the Duma, and most likely his connection with Rasputin would be revealed. No minister with such a record could live with the Duma. Second, he saw at once that he did not need the *starets*, for he had made personal contact with the empress and had won her complete trust. Rasputin was no longer useful as an intermediary, and the dangers of further trafficking with him outweighed whatever benefits might be involved. Rasputin might divulge many things that could ruin him. In a flash he saw the resolution of his problem: murder the *starets*. The killing would destroy the evidence of his use of Rasputin, and would make him a hero among persons fo a wide range of political views, per-

---

15. Alexandra also feared for the lives of the imperial family. She told Khvostov that his appointment depended on his ability to protect Rasputin and her family. "Imperator Nikolai II i ego pravitelstva," *Russkaia letopis*, II (1922), 17. She further informed him that "You will be appointed, but only on the condition of taking Beletsky as assistant minister. He is the only person who can protect us."

16. The story of Rasputin's dabbling in church affairs is in Curtiss, pp. 388-409.

sons who were coming to believe that the political salvation of the country was in the destruction of this man.

However, disaster began when the ministers plotted murder. Rasputin did little to bring about their their ruin—they destroyed themselves. Each feared betrayal by the other; as a result they called in persons to assist in the plot.[17] Inevitably, word got out about what they were doing, and when it reached the ears of the emperor late in February of 1916, they were fired. Although Nicholas knew of the homicidal intentions of Khvostov toward Rasputin, he talked of bringing him back into the government when he had matured.

The collapse of the entire affair left the other participants quiet for the moment. Alix was chastened and silent. Rasputin left discreetly for a month's stay in Siberia. Khvostov was ordered out of the capital, and he left without protest. Beletsky, who had a compulsion to confess, published in a St. Petersburg journal a short version of his plotting. However, his revelations did not save him. He lost a governorship, narrowly saved his seat in the Senate, and was forced into exile from the capital.[18]

§

At the end of 1915 I.L. Goremykin, the premier, indicated that he was looking with trepidation to the forthcoming meeting of the Duma. He was tired of the responsibility of power and yearned for dismissal. Rodzianko had taken the unprecedented step of writing him a personal letter urging him to resign. The Duma now did not tolerate him at all; every mention of his name and every appearance he made caused united cries of "resign!" to be hurled at him. In 1915 the emperor mentioned to Alexandra that he would soon let the old man go. He did not indicate when this would happen or who he was considering as the next Chairman. Alexandra agreed that Goremykin was a liability and should be released from his burden. Beyond that she said nothing; her husband had not confided in her the immensity of the dismissal.

At this moment the emperor was taking a long view of the national crisis; more and more he had been losing hope of finding some kind of political stability under the existing system. Nevertheless, it was necessary to harmonize political life so that the country might unite to win

---

17. Trufanoff, *Monk*, p. 197.
18. Spiridovitch, *Raspoutine*, pp. 304, 315.

the war. The generals were telling the tsar that politics and its dissensions were undermining the nation's will to fight. There seemed to be only one way out: the creation of a dictatorship under two leaders, one for the civilian sector of national life and another for the military. Although there was no agreement among the military chiefs concerning the details of the plan, which was still in the stage of early discussion, General M.V. Alexeev was widely supported when he submitted the name of the Grand Duke Sergei Mikhailovich for the role of military dictator.[19] It was expected that the tsar would provide the name of the political leader, and it was hoped that with this arrangement all political problems would remain frozen by a truce until the end of the war. Nicholas slowly worked out the criteria that he would use in selecting his man; in the meantime Goremykin held his post with increasing fear and impatience.

Nicholas might have tolerated a progressive Chairman of the Council, he believed that he needed a person of unquestioned loyalty, since a man of constitutionalist leanings might not be able to resist the temptation to use his powers to weaken the autocracy. Loyalty was far more important than ability, because the monarchy would be risking its existence in the hands of the new prime minister. Nicholas decided to move cautiously; he planned to appoint someone to the position without revealing that he was being observed so that his ability for the post of dictator might be judged. With these thoughts in his mind, the emperor began to cast about, seeking his man.

In the last few days of 1915, he was considering Boris Stürmer.[20] Stürmer was a well known and widely hated reactionary who had in an earlier generation won fame by crushing the liberal elements in the local government of Tver Province. Now almost seventy years old, he had acquired considerable experience in a variety of major bureaucratic posts including governorships, positions in ministries, and a seat on the Council of Empire, where he was hailed as a pillar of the ultra-conservative wing. He had demonstrated that firmness, ambition, and deceit were hallmarks of his character.[21] In the atmosphere of the years after 1905 Stürmer was too notorious and too closely identified with the worst days of tsarism to be appointed to any post as long as the government was interested in avoiding violent controversy. In January, 1916 avoidance of controversy was still an important matter, but other considerations had risen to the fore. Nicholas believed that a strong man was needed who

---

19. Semennikov, *Monarkhiia*, pp. 259-266.
20. *PTR*, Stürmer, I, 221-92, V, 159-89; Manasevich-Manuilov, I, 361-402; II, 9-23.
21. Pares, *Fall*, p. 318.

could control the Duma even if its temper was inflamed by his presence. Stürmer's record showed that he was loyal, and it also revealed that he had successful experience in handling political bodies hostile to the regime.

In the Council of Empire he had made no secret of his views, and he publicly asserted that he was ready to undertake any difficult or unpopular tasks on behalf of the monarchy. He was the only major political figure eager to serve as Chairman of the Council; there were no other volunteers. In his own salon, which met weekly and included many of the reactionaries in the government, he proclaimed his readiness for office.

He was the only person interviewed for the position. The chairmanship was a post into which a candidate could not be lured or duped at this time; he had to enter the cabinet willingly. In their talk, the tsar wanted to know if Stürmer would deal firmly with the deputies when they became troublesome. In answer to this query, Stürmer gave assurances of his toughness. Nicholas also inquired if he had any program. Stürmer denied that he had, he said he would do what he was told. This was the correct answer. The tsar wanted no political programs, changes or reforms of any kind, either from the cabinet or from the Duma. For the moment he wanted tranquility. It was his intention to hold on to power, and this could be done only if the Duma and various public organizations were prohibited from instituting changes. The emperor was hoping that he could enforce a political truce.[22] Stürmer, in other words, had guessed what Nicholas wanted and had given two correct answers. Alix, Metropolitan Pitirim, Vyrubova, Rasputin and others pushed the cause of Stürmer but most likely it was his own performance before the tsar that won him the post.

During the interview Nicholas did not reveal his plans for a future dictatorship. It was not hard to imagine the resistance such an idea would encounter. Alexandra was certain to be in the foreground of the opposition. For personal reasons she disliked the Grand Duke Sergei Mikhailovich. She also hated Alexetev, the originator of the plan, who was regarded by her as the closest military ally of Guchkov. For a while she had insisted that the two were corresponding and plotting, but the tsar had finally put an end to her accusations with a few petulant phrases denying the allegation. But she would not be the only opponent. The Duma would, of course, have much to say. In fact, when Rodzianko eventually heard about the plan from General Manikovsky, who was against it, he quickly went on record to oppose the dictatorship and he threatened to make it a

---

22. Spiridovitch, *Raspoutine*, p. 283. Paléologue, *Memoirs*, II, 169. Rodzianko, p. 171.

subject of debate in the Duma.

In spite of the tsar's secrecy the news that Goremykin was about to step down was widely known. One of the most experienced political onlookers was I.F. Manasevich-Manuilov, who had spent an adventurous lifetime in the subterranean world of tsarist politics, police and journalism. A curious figure who combined low instincts and occasional lofty behavior—after the revolution he was one of the few who remained loyal to the imperial family—Manasevich-Manuilov had ranged widely in the service of the police. He had served as a successful spy and briber in the Russian diplomatic missions and later he had moved to Paris where he was active in special campaigns aimed at bribing the French press in order to secure favorable publicity for the Russian-French treaty negotiations.

In the decade before 1918 he had cultivated the journalistic side of his career and had been serving for some time as a correspondent of the *New Times*. This job had given him some understanding of the realities of domestic politics. He had recently watched with admiration what he thought was the important role that Prince Andronnikov, with the help of Rasputin, had played in assisting A.N. Khvostov to power, and came to the conclusion that after the fall of the prince he too might help in elevating some suitable bureaucrat to high office. From the numerous vantage points he had established, Manasevich-Manuilov heard the rumors of Goremykin's impending departure and he thought that this offered him the opportunity to undertake a new life's work as the maker of ministers. He noted Stürmer's eagerness for the job so he went to him and offered his services. Stürmer knew him—they had met in Rome—and he knew the man's reputation for achievement, so he accepted the offer.

Stürmer played his part by saying the right things to the tsar. At the same time Manasevich-Manuilov launched his operations in the underworld of intrigue. He believed that Rasputin was all-powerful and that his influence made and unmade cabinet members. This belief was commonly accepted among the journalists with whom Manasevich-Manuilov associated. He thought that it was his job to get close to Rasputin, convince him that Stürmer was a good man, and then induce him to lend his support to the candidate. Unfortunately, he had written several articles denouncing the *starets* and these had been published in his paper. In addition, Rasputin did not like Stürmer.[23] The journalist was aware of these hurdles, but there were others of which he was not.

---

23. Eugene de Schelking (Shelking), *Recollections of a Russian Diplomat* (New York: The Macmillan Co., 1918), p. 117.

Alexander Khvostov, the uncle of Alexei, had warned Nicholas that Stürmer was a doddering old schemer. The empress at first pushed the cause of the younger Khvostov for the post of prime minister, despite the scoffing of the emperor. Later, however, Stürmer won part of her sympathy by expressing a friendly feeling for Rasputin, who did not want Khvostov in the Chairmanship.[24] There he might become too powerful and independent. To Rasputin, Stürmer seemed to be the perfect archetype of the landlord in government; he suspected that the man was not to be trusted, that behind the facade of fawning obsequiousness was a certain degree of independence.[25] He also noted that Stürmer's clumsy attempt to change his name to Panin served only to draw attention to one of his worst failings—his German name. Nevertheless, he might be less dangerous than the immensely ambitious Khvostov, so the *starets* was willing to pronounce a tepid blessing over Stürmer. At the same time he left the impression that he would be ready to withdraw this blessing anytime before the nomination was made.

He was struggling to achieve the same degree of detachment and indifference to Stürmer's cause that the empress had. Since neither of them were happy with the candidate, and since the tsar had from the outset indicated that his mind was almost made up, they could only cooly agree with him. The empress sought a compromise and suggested that Stürmer might be given an interim appointment. Then when the Duma was quieted he could be dismissed and Khvostov could move to his place. Inasmuch as the emperor wanted to observe Stürmer at a distance in order to estimate his ability to fill the position of dictator, he did not object to this idea.

Manasevich-Manuilov knew nothing of all these reservations on the part of Alexandra and Rasputin. But the barriers he did know of did not stop him. His knowledge of backstairs intrigue and his wide circle of acquaintances could assist him. His instincts in these affairs served him well. He was acquainted with Nikitina, one of the maids of honor at court, whose father was commander of the Peter and Paul Fortress. She was a devotee of Rasputin and was in a position to recommend Manasevich-Manuilov to him and to help Stürmer and Rasputin to meet secretly.[26] He also knew the gadabout Osipenko, the secretary of Metropolitan Pitirim, who was close to Rasputin. Osipenko handled much of the business that Rasputin transacted with the metropolitan, and he frequently took the

---

24. A to N, January 7, 1916.
25. Rudnev, p. 47.
26. Volkov, p. 74.

Rasputin children to one of the circuses in the city. Manasevich-Manuilov told Osipenko of his desire to meet the *starets* and of his newfound admiration of him. He was then brought to Pitirim who was to decide on the sincerity of his words; Pitirim announced that his conversion to the cause of the *starets* was genuine. Finally, he met Rasputin and began to pay frequent visits to his apartments. Through machinations he managed to secure the use of a military automobile which both men used to ride about the city. Rasputin was much impressed by this gesture because the car was large and powerful and could out-speed any civilian car. He had feared that he might be assassinated by being pursued during one of his jaunts to some lonely part of the city, but with the new car he felt safe. When A.A. Polivanov, the Minister of War, heard about the vehicle he ordered it returned at once to the army.[27]

By such small gestures of good will Manasevich-Manuilov won some of Rasputin's trust. The *starets* was not one to discourage anyone who thought that he was powerful in the government. He in turn used his new companion who he knew could spread the word of the Rasputin power among newspapermen. He was careful to expose Manasevich-Manuilov to the same treatment that had impressed Iliodor. All his actions were a caricature of what he imagined those of the busy man of affairs to be. He hastily dashed off memos and orders to important people, many telephone calls were made to and received from the palace, there were conferences at night, hints of imaginary consultations with ministers, boasts of how he spoke his mind to the tsar, and all of the other paraphernalia with which Rasputin surrounded himself when he wanted to impress someone with his exalted position. He built a shadow ministry which existed mostly in the realm of make-believe.

The object of all this showmanship was duly impressed and he, of course, passed the word of what he had seen to all the circles of his friends and acquaintances. While the appointment of Stürmer was under consideration by the emperor, Rasputin could do little to hurry action. He behaved like a man deep in thought, a man judiciously turning over in his mind important matters of state because the responsibility of action rested with him. He promised careful consideration of the important question of the nomination of Manasevich-Manuilov's client, but of course he had already made up his mind. The journalist was grateful and was convinced that he was making headway with the one person who was

---

27. Alfred Knox, *With the Russian Army, 1914-1917* (London: Hutchinson & Co., 1921) II, 412. *PTR*, Delivanov, VII, 78-79.

important in securing nominations.

Because Nicholas did not confide in his wife at this time, we do not know about the stages through which his thought progressed as he sought to reach a conclusion about the candidate. In mid-January 1916 Rasputin rightly concluded that Stürmer would certainly be named chairman.[28] Although Rasputin realized that he could have no effect on the course of events he did not want Stürmer to know this. He moved cautiously and claimed that he was actively supporting the nomination. In order to impress his friends he asked Pitirim to push the cause of Stürmer during an interview with the tsar. Since the emperor's mind at this time appeared to be made up—Rasputin concluded that his expression of his opinion could offer no offense. At the interview Pitirim, however, first chose to mention the Duma, presuming that this would permit him to go on to the subject of Stürmer. He thought that this would give him the opportunity to remind Nicholas that Stürmer was the one man who could tame the Duma. His remarks were cut short at once. The emperor deeply resented the metropolitan's interference in politics and he ended the interview. That night he wrote an angry letter to his wife, inquiring about Pitirim rhetorically, "I should like to know who influenced him in this matter?" Neither Alexandra nor Rasputin mentioned the nomination again. Fortunately for Rasputin, Manasevich-Manuilov did not hear of the incident.

Although Stürmer represented himself at court as the man of steel in dealing with the Duma, he actually saw no reason to arouse needless hostility among the deputies. He knew that the old days were gone; no longer could a servant of the tsar accomplish all things as long as the imperial will supported him. He believed that when the Duma was in a mood to unleash strong attacks no chairman could stay in office. Six days before his appointment he began to mend his political fences in the hope that he might induce the Duma to greet him with less belligerency than might be expected. This act involved him in a betrayal of the agreement with the tsar; nevertheless he pushed ahead, requesting that Manasevich-Manuilov make overtures to Rodzianko. This risky experiment had little hope of success, but it had to be made. The journalist went to Pitirim and asked him to carry out the unpleasant assignment. The metropolitan revealed to the president of the Duma that Goremykin would soon go and that there was some reason to expect that Stürmer would succeed him.[29]

---

28. N to A, January 5, 12 and March 5, 1916.
29. Rodzianko, p. 172.

He added that if Stürmer received the post he wanted friendly and cooperative relations with the Duma. Rodzianko considered this visitation one more piece of evidence that since Rasputin and his gang knew about ministerial changes in advance they must therefore have had something to do with those changes. He refused to make any promises to Pitirim but lectured him on the need to disassociate himself from Rasputin.

On 20 January the emperor from the Stavka announced Stürmer's appointment. Stürmer at once began to look for support. He realized that Rodzianko's threatening stance typified the feeling of the majority in the Duma, so he turned to the ministers and certain other well-known figures in the capital. It was his plan to win promises of cooperation by vowing that he would try to work with the best elements in society. Most of the persons he was able to talk to were uninterested or non-committal. For instance, he visited A.N. Naumov, the Minister of Agriculture, in an attempt to enlist his support. Naumov revealed that he wanted to make enemies of no one, and then he went on with the lugubrious tale of his own appointment. He complained that he did not want the post, did not understand why he was appointed, and hoped that Stürmer would show mercy by dismissing him at the first opportunity. It is hard to say if Stürmer's tactic achieved anything. Political tensions continued throughout most of his tenure as prime minister. However, the political storm that many onlookers anticipated would come the moment he took office was long delayed.

Manasevich-Manuilov also busily sought to make allies for his client. He made an unsuccessful visit to the French ambassador, Maurice Paléologue.[30] The two had met years before in Paris and the ambassador was well aware of the evil reputation of his visitor. However, Manasevich-Manuilov knew that his status as a correspondent of an important newspaper would force Paléologue to see him and to be outwardly cordial to him. It was Stürmer's conviction that he eventually would become Minister of Foreign Affairs. In this post S.D. Sazonov had enjoyed relative immunity from attack either by the tsar or the Duma in spite of the fact that he was politically active. Stürmer assumed that this ministry was a kind of sanctuary and for that reason he wanted it. Ministers who dealt with allied diplomats were usually treated well by the Duma. the task of Manasevich-Manuilov was to try to create the impression that Stürmer would work with the allied diplomats. Although Manasevich-Manuilov offered to take any requests of Paléologue to Stürmer, the ambassador refused the offer.

---

30. Paléologue, *Memoirs*, II, 168-69.

Stürmer had involved himself in a dangerous maneuver: he had promised the emperor that he would silence any outburst by the Duma, but on the other hand he tried to spread the word that he would attempt to work with the Duma. For a while the lack of large-scale political crisis hid his double game because he did not have to make a choice as to which pledge he would redeem. Only gradually did the emperor come to suspect the man's weakness. When he chided him for being over-conciliatory in some of his dealings with the Duma, Stürmer put him off. He said that he was experimenting with ways to trick the deputies into acquiescence. At the same time he indulged in numerous gestures of goodwill, flattering Nicholas by saying that he was a new Tsar Alexis and it was his aspiration to be his Ordyn-Nashchokin.[31]

While the emperor suffered only a slow disenchantment, Rasputin experienced a swift one. Shortly after the appointment, Stürmer personally thanked Rasputin for his assistance. However, the *starets* heard about the long and intimate conversation the new chairman had with Naumov and several other persons in the government, and he correctly surmised what Stürmer's game was. For proof, he sought a meeting with Naumov, but was refused. Desperate for information that he could use in a confrontation with Stürmer, Rasputin sought to force his way into an interview with the Minister of Agriculture, but he was not successful. From other sources Rasputin gathered facts that convinced him that Stürmer meant to go back on his agreement about dealing with the Duma, and that he was also seeking support that would make him independent of Rasputin. Eight or ten days after the appointment was made the two men had a noisy quarrel. Manasevich-Manuilov, who witnessed it, reported that loud words and denunciations were exchanged. Finally, Rasputin warned, "You ought not to oppose the plans of Mama. Pay attention. If I leave you, take care." But he was in no position to carry out the threat. Only a few days before he had added his own esoteric approval to the voices which were acclaiming Stürmer as the best man for the post. To go suddenly to the tsaritsa with the news that he had so soon changed his views might have hurt his reputation for infallibility. True, his assent had not been given with enthusiasm, but only he was aware of his reservations. He recalled the strong reaction of the emperor to Pitirim's attempt to interfere with this appointment, and it was probable that any attempt to bring about a dismissal would be equally unwelcome.

---

31. A. Nekliudov, *Diplomatic Reminiscences Before and During the World War, 1911-1917* (London: J. Murray, 1920), p. 410. Afanasy Laurentievich Ordin-Nashchokin (1605-1680) was a historically influential and broadly admired statesman.

To add to Rasputin's frustration, Stürmer's policy in dealing with him was approximately correct, for the wily old bureaucratic had, like Goremykin, come close to guessing the truth about Rasputin's so-called power. Stürmer knew that Rasputin's support was a minor ingredient in the appointment process and that it had to be employed sparingly even though it was useful to the candidate seeking to use every way to ease his path to power. He understood that Rasputin was more useful before an appointment than after. Once Stürmer was in office and there was less need for the *starets*, he adjusted his policy toward him accordingly, because Rasputin was a liability for him when he had to deal with the Duma or with many of the ministers.

Now the shoe was on the other foot, and Rasputin found that he was in the anomalous position of being forced to support Stürmer in order not to discredit himself. The Duma was about to convene, and there was fear at court that it would take this opportunity to express its displeasure with the new Chairman of the Council. Stürmer had done what he was able to do to head this off, but no one thought that he had accomplished much. Alexandra busily searched for a solution; she was the only one who had faith that there was a solution. She recalled that before the opening of the previous session Rasputin had suggested, in order to save Goremykin much pain, that the Tsar appear in person at the opening ceremony. This was supposed to make it impossible for the deputies to hiss the old man. The empress now revived this notion in order to afford some protection to the new chairman. Rasputin was chagrined to see that his idea was to be used in a new situation where it was of no value to him, but he could not avoid expressing his approval because Alexandra was much taken up with the idea as a pat solution to the current crisis. As usual for her, his assent meant that the plan had almost divine endorsement.[32]

Nicholas accepted the plan and offered his ceremonious presence. The little speech he delivered was highly conciliatory in tone; it created a minor sensation because in it he greeted the assembled delegates as the representatives of the nation. These words plus Nicholas' promise to Rodzianko to seriously consider a Ministry of Confidence annoyed the empress but did much to quiet the assembly.[33] Even many of those who did not take it seriously or even those indifferent to its significance felt that it would be imprudent to begin a policy of baiting the government after this rare instance of recognition of the Duma's essential role. When

---

32. Pares, *Memoirs*, p. 366. Rodzianko, pp. 174, 176. Maurice Verstraete, *Mes cahiers russes* (Paris, 1920), pp. 45, 47. A to N, November 13, 1915.

33. *PTR*, Rodzianko, VII, 130.

the tsar had finished addressing the assembly, Stürmer rose to deliver his first greeting. It was a dull talk couched in the profoundest kind of bureaucratese, and it was received in silence. Only V.M. Purishkevich, the arch-conservative, rose to show any sign of belligerence.

The general atmosphere of a truce that resulted from the events of the opening ceremony did not benefit Rasputin. Stürmer boasted that it was he who had succeeded in inducing the Duma to follow peaceful ways and for the moment there was little that Rasputin could do to prove that this was not true. As long as the political skies seemed clear for the while, Stürmer had not much need for Rasputin.

By the summer of 1916 Stürmer had gathered in several more reins of power and these made him more independent than he had been. Shortly after he became chairman he took over the Ministry of Interior when Khvostov was fired, and in July he relinquished that post and stepped up to the Ministry of Foreign Affairs. He now had great formal power inasmuch as he held two ministerial portfolios.

Stürmer's tenure of office was a time of unrealized hopes and humiliations for Rasputin; his activities were confided to small deals and to participation in some obscure appointments.[34] Stürmer had no need to grant him any more favors, and indeed, Rasputin hesitated to ask for more lest the real nature of his power be revealed. Alexandra freely passed on to her husband Rasputin's remarks and observations on many affairs close to his heart. The emperor heard these but said nothing.

At this time the empress had been chastened too; she was shocked to learn that her husband had been considering temporary change to a system of dictatorship. Moreover, her ignorance of events was revealed to Stürmer since he had gone to her when first he received the rumor of the proposal, and she had assured him that no such step could be under consideration without her knowing about it. Nicholas by this time had lost faith in the scheme, but Alexandra had already been compromised.

Stürmer felt more secure and independent than before. Russia had unleashed its successful offensive in the late spring of 1916, and critics of the government were for the moment quieted. Therefore, Stürmer could afford to be cool toward the *starets*. They seldom met and the chairman even ignored many of his messages. Now that he had reached a new plateau of power he was in a mood to divest himself of all his old, hindering alliances. He suddenly decided to get rid of Manasevich-Manuilov by casting him out in disgrace. But he involved himself in a dangerous

---

34. Rudnev, p. 40.

maneuver when he tried to pull away from his assistant and from Rasputin together. If a public uproar resulted it might eventually lead to an unmasking of the former intimate relationship of the trio.

The Okhrana official E.K. Klimovich, a sometime friend of Rasputin, had in July of 1916 come to admire his new chief, the elder Khvostov, newly named Minister of Interior.[35] Khvostov knew much about Manasevich-Manuilov's career as a swindler and he hated him. Klimovich at this time also took note of Stürmer's desire to destroy Manasevich-Manuilov. Consequently, he baited a trap to catch the unsuspecting confidence man in one of his pursuits—blackmail. The victim was caught and arrested.[36] It is hard to discern what were Klimovich's motives and who he was trying to please in all this. Stürmer feared a public trial, so he tried to hold up the investigation conducted by the Ministry of Interior. At the end of the year the tsar had to intervene to order the indictment discontinued.[37] Stürmer fired Klimovich.

Rasputin watched the investigation and its attendant publicity with much apprehension and disapproval. The dismissal occurred while he was in Siberia. It worried him much because it endangered his life. Whenever a high police official left office there was usually a period of confusion in which the police guards who watched Rasputin were replaced by others. In Siberia he was in an especially exposed position: the police complained that it was much more difficult to protect him there. At this time he had to make the dangerous trip back to the capital with a new and inefficient police escort looking after his safety. The incident demonstrated to him who had the upper hand in the Rasputin-Stürmer relationship.

Shortly before Rasputin's death, Stürmer was relieved of office. Although Rasputin approved of the change, he did not cause it. It was brought about by the attacks of the Right and Left in the Duma.[38] Even without any sharp political crises such as those that occurred during most of the second half of 1915, the country had drifted into trouble. It seemed that the war would never end, there was a scarcity of food and fuel in the principal cities, and the nation was without real leadership. The patience of the Duma was at an end. A number of party leaders sensed the growth of dissatisfaction in the country, and at the meetings of the Duma there was a series of attacks against the government. Stürmer, the empress and

---

35. PTR, A.A. Khvostov, V, 456-57; Klimov, I, 57-61.
36. Rodzianko, pp. 211, 213.
37. Semennikov, Politika, p. 122.
38. N to A, June 24, September 20, November 8, 1916. A to N, June 23

Rasputin were accused of treason. Under the first assaults Stürmer crumbled and asked to be relieved of office at once. For some time Nicholas had been displeased with his irresolution, and consequently dismissed him.

§

This situation brought a new and unexpected crisis to the regime. The chairmanship of the council could not be left open for any length of time, yet there was no suitable person willing to fill the post. Almost at once the emperor knew that he was going to have to turn to a man who was merely the least undesirable one of a bad lot. It was important that someone be found who would shut down the Duma before it did any more damage. Surveying his cabinet for a possible candidate, he came upon the Minister of Ways of Communication, A.D. Trepov, who had solved some difficult administrative problems in completing the Archangel-St. Petersburg Railway. Neither Alexandra, Nicholas or Rasputin liked the man.[39]

For Nicholas, the main problem of controlling the Duma was as critical as the quest for a chairman. He was now faced by a menacing coalition of forces more formidable than the moderates of the Progressive Bloc of 1915. The opposition had launched an unprecedented campaign of gossip, containing remarks ranging from sly innuendos to crude fantasies about a court camarilla that under Alix's leadership was preparing a separate peace. In the Duma deputies were cheering violent speeches directed against the government, the empress and Rasputin. This frontal assault no longer dwelt on the ineptitude of the government; the theme was treason. In such a situation the danger was acute, and the emperor was willing to settle for any solution that would deliver the government from the immediate threat. He believed that the appointment of Trepov could fit the desperate need. Trepov promised, in his reluctant acceptance of the post, to make an effort to placate the Duma. But he did not take the job without submitting extensive demands, all of which were painful to the tsar. Nicholas needed the man but could not give him what he wanted. Eventually, Nicholas thought he saw a way out of the dilemma. He decided to hold on to Trepov until the Duma quieted, and then he

---

39. Jacoby, p. 72. A to N, November 1, 1915, November 7, 1916, December 5, 1916.

would fire him. The tsar did not confide to the minister that his tenure was to be temporary. There was a kind of tawdry Machiavellianism about the plan but Nicholas was proud of it, and he confessed to his wife that,

> It is unpleasant to speak to a man one does not like and does not trust, such as Trepov. But first of all it is necessary to find a substitute for him, and then to kick him out—after he has done his dirty work. I mean to make him resign after he has closed the Duma. Let all the responsibility and all the difficulties fall upon the shoulders of his successors.[40]

As a prelude to the dramatic interview in which he offered the post to Trepov, Nicholas prayed before his favorite icon. He heard that the obvious panic of the government in the face of the Duma attacks had emboldened Trepov to make sweeping demands that would have to be met before he accepted office. It would be hard for Nicholas to deny that the new appointment was not an act of last resort. His task was a delicate one: he had to induce Trepov to defer making his demands, to persuade him that they ought not to be conditions for acceptance. If they could be placed on the agenda for discussions to take place after the appointment was made, then Nicholas could by various devices put off Trepov until the time came to dismiss him. Nicholas was successful in this plot, and at the same time he extracted a promise that Trepov would recess the Duma. To his wife, Nicholas revealed only that a vow had been made concerning the Duma. He feared to alarm her by mentioning that Trepov had won one concession: Nicholas had promised to get rid of the three most unpopular ministers—Protopopov, Count A.A. Bobrinsky and Prince V.N. Shakhovskoi. Nicholas did not regard any of these very highly and would have been willing to send them away if Trepov had not made the request, but the empress by this time regarded Protopopov as the lone defender of autocracy in the cabinet.

Since Trepov hated Rasputin, Alexandra had no difficulty making up her mind about the new Prime Minister. The next two appointments could not please her. The incompetent Bobrinsky was fired from the Ministry of Agriculture and replaced by the capable Rittikh who had associated with Stolypin. Another excellent man, Pokrovsky, was rumored to be under consideration for Minister of Foreign Affairs. It must have seemed to Alix as if Trepov was naming his own cabinet. She announced that he

---

40. N to A, December 4, 1916.

was a revolutionary. His actions during the next few weeks did nothing to change her mind, and soon she was stating that she would like to see him hanged. Nicholas retorted with an inquiry as to who she wanted in the post. Protopopov was her choice, but the emperor was not ready to hand over the cabinet to a man whose sanity he questioned. Alexandra next offered a few desperate suggestions. She thought that Trepov might be made chairman of the Council in name, while Protopopov served as the real leader. Both Nicholas and Trepov rejected the plan, which she would not have made had her husband confided to her the ambition and stubbornness of Trepov. Then she asked that Trepov be named a temporary chairman who could be dismissed once the crisis was past. This formulation was closer to the trend of the emperor's thought and was therefore acceptable.

Throughout his tenure Trepov was pursued by the angry empress. Because the crisis in the Council of Ministers had broken without warning she had been unable to supply her husband with the names of any candidates for the chairmanship. Rasputin had failed her since his endorsement of the Minister of Interior, Protopopov was lukewarm, and her attempts had failed to get him to say that the appearance of Protopopov was providential. As a result, it had been impossible to hide the fact that the naming of Trepov had been an act of last resort. The difficult position of the Tsar was barred.

Nicholas's sense of resignation and pessimism grew deeper as the days passed. He spent the last two weeks of November at Tsarskoe Selo. After these vacations he usually returned to the Stavka in a mood of refreshed hope. But at this moment he left full of gloom, and Alexandra, knowing of plots against the government, and fearing that he might abdicate or make some other desperate move, sent a written sermon after him.[41] The letter brimmed with promises, warnings and demands, exhorting him to stand firm. She used every theme in her repertoire to bouy him up.[42] In addition she mentioned Rasputin and noted how his presence and support was a sign of divine assistance to come. This interpretation meant more to her than it did for Nicholas. His state of depression continued as he oscillated between seeking occasional help and advice and rejecting it when it was forthcoming from her. He reminded her that she was talking to him as if he were a child. She adjured him to "be Peter the Great, Ivan the Terrible, Emperor Paul—crush them all under you."

---

41. A to N, November 5, 1916. "It's time for the saving of your country and your child's throne...You are an annointed one."
42. Simanovich, p. 167.

Once Trepov took over the chairmanship of the council, her demands and pleas to her husband increased. She briefly outlined her program: it included large-scale deportations of the Duma leaders to Siberia. She accused the new chairman of holding talks with Rodzianko, and she insisted that they be discontinued at once before they led to seditious acts. When her charge was proved wrong she simply went on to other matters. At an unabated pace the suggestions for policies and nominations went on. Nicholas occasionally considered her importunings seriously, but for the most part he ignored what she said. She was now almost hysterical with fear that he would crack. She was crushed by the news about the disposition of the Duma. For some time she had been insisting that it be completely dissolved on 27 November. There was a rumor that the Grand Duke Nicholas would be in the city on 30 December, and she feared that demonstrations centered on the Duma were going to hail him as a new tsar. In addition, Rasputin warned that it would be dangerous to permit the deputies to stay in the capital during a period of recess. But the empress was one of the last persons, as she complained, to hear about the recess. Even Rodzianko knew of it before she did. She petulantly accused her husband of listening to Trepov instead of to her.

When Nicholas first announced that Trepov was under consideration he wrote to Alexandra asking her not to bring Rasputin into the question of appointments; he wanted to decide the matter alone. She chafed at the restriction and could not resist pointing out that the proposed minister was a deadly foe of the *starets*. Nicholas knew that she believed that the state would not flourish as long as foes of Rasputin were in positions of importance. But he ignored her warning. Rasputin stayed clear of the nominating process, but he was vitally interested in it and had much at stake.

During the last few weeks of his life, Rasputin's moves were mostly designed to secure protection against assassination. He was obsessed with dread and often mentioned to his daughters the danger of murder. He talked about his will and what would happen to his family if he were killed. Secondly, he was interested in preserving his position at court. As he watched the tsar respond to the crisis from the middle of November he became depressed. He feared that the emperor was not dealing firmly enough with trepov, who had begun his ministry by asking that a number of appointments and dismissals be made. Under these requests the only ministers whom Rasputin regarded as trustworthy, Bobrinsky, Shakhovskoi and Protopopov were marked for removal. Rasputin did not know the tsar's game, and when the request of the chairman was not flatly rejected he suspected that the weak Nicholas was eventually going to give in. Rasputin was pleading for the nomination of Kurlov as Okhrana chief, but nothing came of it. He regarded this as another example of Trepov's power. Rasputin centered his complaints on the issue of Protopopov's tenure. He had no illusions about Protopopov, but he believed that he was the only brake to Trepov's drive for real, independent power. Protopopov's position was already insecure, and Nicholas was preparing to fire him.[43] To Rasputin these facts plus the known pessimism of the emperor indicated that he was tiring of the effort to hold power. Whatever was to be the outcome of all these issues, Rasputin saw that his own position was not secure.

He saw a great danger in the Duma. It had been heartened by the realization that its sensational attacks on the regime in the autumn of 1916 had attracted the attention of the country. When the government tried to divert attention by revealing its diplomatic triumphs in the London agreements (in which the tsar was promised control of the Bosphorus) the effort failed. Rasputin believed that if the Duma continued to sit it would become bolder. Nicholas had been forced to settle for a recess instead of a dismissal; in his view to have asked for more would have been to ask the impossible. But Rasputin warned that if the deputies were dismissed for three weeks only, they would not return to their constituencies but would stay in the capital to talk and consult—and possibly to plot. He understood that the members of the Duma were important and strong only when they were together. In their home districts many of them were unknown. Because of this he saw that the recess might be more dangerous than having a Duma in session. The empress several times passed on this advice to her husband but he made no response.

Rasputin accepted the appointment of Trepov as stoically as possible. But if he accepted Trepov, Trepov did not accept him. The minister's conditions of acceptance of the chairmanship contained a hint that the problem of the Duma would be greatly simplified if the *starets* were to leave. When political negotiations failed to bring any solution to the problem, Trepov chose to deal directly with Rasputin in the hope that he could get him to leave. This coup would then set the stage for the removal of Protopopov. His proposed solution was direct, simple, and dangerous. He went to his brother-in-law, A.A. Mosolov, Head of the Court Chancery, and asked him to make an offer to Rasputin, promising certain gifts

---

43. N to A, November 10, 1916.

if he would go away.[44] Then began the dealings between the minister and Rasputin.

Trepov underestimated the difficulty of disposing of the *starets*. He knew Rasputin only through the gossip in the city and in the newspaper stories that alluded to him. Using these facts for judging the man, he concluded that he was involved with a person interested in money and religion. He offered both to Rasputin in the form of a 200,000 ruble bribe plus the curious promise that he would also receive a church. He apparently thought Rasputin was a priest. These were to be given when Rasputin withdrew to Siberia. Mosolov reluctantly carried out his part of the scheme, although he first pointed out that if Rasputin were to take the story of the bribe to Tsarskoe Selo there could be much trouble for all persons involved in the plot. Trepov replied that the possible gain was worth the chance. Desperate times demanded desperate risks.

However, Rasputin spurned the offer. Because his only goal was to protect Protopopov's position he did not make an issue of the bribe. Destroying Trepov might not help his cause. There was always the danger that the next occupant of the chairmanship might be even more hostile to him. In addition, he could not recommend any other person for the chairmanship. The firing of Trepov amidst such unsettled conditions might make the problem of Rasputin's security even more difficult. In order to put Trepov in his place, Rasputin sought to intimidate him by saying that he would soon depart for Siberia, and the empress in alarm would call him thence with promises of increased power. He said that if Protopopov was fired he would see to it that Trepov was fired too. The minister ignored the threat and continued to demand of the tsar the dismissal of Protopopov because there could be no peace with the Duma as long as he was in office. Rasputin's reply to this claim was weak; he continued to remind the empress to tell her husband that the life of the dynasty depended on Protopopov's tenure of office. Here the duel between the *starets* and Trepov ended when Rasputin was murdered at the end of December, 1916.

§

---

44. Mosolov, pp. 170-173.

In September, 1916 there were street disturbances in St Petersburg and Moscow. Long lines of people waiting for food or fuel were often disappointed at the stores when supplies gave out. At first there were peaceful demonstrations to protest the shortages. However, the situation worsened when winter approached and as a result, the demonstrations sometimes became riotous. The police were tirelessly warning the government that the lack of supplies was creating the kind of mass discontent that could lead to revolution.[45] Rumors flew about that when the Duma met in mid-November it would seek to use this discontent to overthrow the government. Once again the regime had to seek a strong man who had some ideas about managing the Duma and who could use the Ministry of Interior to deal with the riots. No volunteers stepped forward; from the ranks of the bureaucrats and court officials there was silence. In the capital gossip reported that the tsar was looking for a moderate member of the Duma to fill the post. It was going to be necessary to find a deputy who would have the confidence of the Duma. The tsar, in other words, had reached an impasse and would be forced to make a concession to the demands of the Progressive Bloc, which called for ministers who had the confidence of the nation.[46]

Among the persons who heard these rumors was Peter Badmaev, who had been the friend of Emperor Alexander III. For years he had been a great foe of Rasputin.[47] During his lean years he conducted a clinic where he practiced the naturalistic Tibetan school of medicine.[48] To his establishment came many persons, some of them well-known members of society and government. In the spring of 1915 he began to wonder if there was not some way of restoring his former fame. He had treated the Octobrist deputy and vice-chairman of the Duma, Protopopov, who then became completely devoted to him. A year later Badmaev had tried to help the imprisoned former Minister of War Sukhomlinov by writing on his behalf directly to the tsar.[49] He then launched a cautious attempt at a

---

45. "Politicheskoe polozhenie Rossii nakanune fevralskoi revoliutsii v zhandarmskom osveshchenii," *KA*, XVII (1926), 28.

46. *PTR*, Protopopov I, 111-181, II, 1-8, 148-155, 273-219, IV, 1-116, V, 238-244, 267-285; Vyrubova, III, 233-254.

47. Rodzianko, p. 35. "Pokazaniia A.D. Protopopova," *KA*, IX (1925), 133-155.

48. This school was outmoded by western, scientific medicine, but its practitioners were not guilty of quackery—a charge often used against Badmaev. See Cyrill von Korvin-Krasinski, *Die Tibetesche Medizinphilosophie. Der Mensch als Mikrokosmos* (Zurich, 1953). Pages xxvi and xxvii have a bibliography of works on Tibetan medicine including studies by Badmaev published in St. Petersburg between 1898 and 1903.

49. V.P. Semennikov (ed.), *Za kulisami tsarizma: arkhiv tibetskogo vracha Badmaev*

rapprochement with Rasputin who he suspected of having great political power. He thought that with the help of Rasputin he might be able to perform some service for Alexander Protopopov.

Rasputin accepted his overtures with considerable caution. He talked to the colorful Mongol and enjoyed his company but he kept his distance. Badmaev was busily attempting to point out the good traits of Propotov, noting especially that he longed to serve the tsar in some high office. Rasputin and Protopopov had met casually in the past and had struck up an acquaintance. Rasputin had no liking for the Duma—the landowner's club, as he called it—or its members, Right or Left. By the summer of 1916 he was beginning to learn, under the careful tutelage of Badmaev, that there were many different parts to the Duma and that a large number of its members were monarchists. He professed to dislike the extremists, especially those of the Right, where Khvostov had been. He suspected that he might find friends in the Center. Badmaev revealed to him that this was the land of Protopopov. Then Rasputin had to be convinced that the Octobrist party had two wings and that Protopopov was not in the same group with Rodzianko. But the *starets* was still suspicious. "This Chinese would sell me for a mouthful of bread," he said.[50] When Badmaev introduced Protopopov to him, Rasputin insisted that their meetings take place at his apartment where three policemen stood near the door. The fact that Protopopov cheerfully accepted the invitation impressed the *starets* favorably, because most of the members of the government who had any business with him tried to hide it. Nevertheless, Rasputin did not hesitate to state his reservations about the visitors. He reminded them of the incident that had occured in May 1916, while Protopopov was in Rome as one of the leaders of the Duma delegation visiting the parliaments of the Allied powers. There he had said that all Russia looked forward to the day when there would be a revolution and the Grand Duke Nicholas would sit on the throne. Rasputin also pointed out that in accepting a cabinet post directly from the tsar, Protopopov would, in a sense, be betraying his comrades. He asked what might prevent him from betraying the tsar eventually.

There is no record that the first accusation—about enthroning the grand duke—was ever again the problem for Protopopov in his dealings with Rasputin. There seems to be only one explanation as to why this important fact failed to alarm Rasputin, who had an abiding mistrust of the duke. He knew that Protopopov had for many years been suffer-

---

(Leningrad, 1925), p. 25.

50. Spiridovitch, *Raspoutine*, p. 325.

ing from recurring fits of insanity. His fellow deputy, A.I. Shingarev, who was a medical doctor, had some time ago advised him to seek competent treatments from a physician, but Protopopov feared that this would become a subject of public gossip and ruin his career.[51] Eventually, he discovered Badmaev's private clinic where he was kept for several months in complete secrecy. The treatments he received there seemed to restore his mental stability. Probably all of these facts were told to Rasputin. Only in this way could his suspicions be disarmed. However, when one fear was allayed, another rose up. Rasputin wondered if the man could be trustworthy since he still had moments of madness. He hesitated to recommend him for any post in government even though he knew what Badmaev did not know: that his word of approval alone could not raise Protopopov to a position on the Council of Ministers.

Protopopov had many people working on his behalf. In the Duma he had been regarded as a pleasant and capable deputy whose outbursts of peculiar conduct were merely eccentricities. Rodzianko in June 1916 recommended him to the tsar as Minister of Ways of Communication. Rodzianko had concocted one of his plots; he was saying in the Duma that although the tsar could not accept any ministry of confidence, he, Rodzianko could induce him to accept a Duma member to serve on the council. It was Rodzianko's hope that this man would do the bidding of the Duma. He was to be a kind of secret agent in the council. The president of the Duma chose Protopopov for this role. Nicholas was distrustful of Rodzianko; although he did not know what the president and "his tovarishch Protopopov" were plotting, he was suspicious.[52]

But it was not long until the tsar was hearing praises of Protopopov from all sides. The Allied press and many statesmen in western Europe hailed him. Even in the spring of 1917, Miliukov, in testifying before the investigating commission of the Provisional Government, admitted that Protopopov's conduct during the visit of the Duma delegation in the West was correct and dignified at all times. King George V of England was so impressed by him that he felt the need to write a letter to his cousin, Nicholas II, lauding Protopopov highly and wondering aloud why a place could not be found for such skill in the higher councils of the Russian regime.

Protopopov had returned from abroad to a mixed reception at home. In Stockholm his sanity had deserted him for a few hours and he

---

51. *PTR*, Shingarev, VII, 35. In the summer of 1916 Protopopov invaded Shingarev's bedroom and began to mouth fantastic schemes.

52. N to A, June 25, 1916.

had met with a German agent in a hotel room.⁵³ This incident created a storm of protest in the newspapers. The tsar demanded to see him so that he might give an account of his highly impolitic behavior. Protopopov was unable to recall the incident in Stockholm. He went to the diplomat A. Nekliudov and asked him to recount the story of the meeting to which Nekliudov had been a witness.⁵⁴ To the press and the tsar, Protopopov gave a reasoned defense of his strange conduct. He also intelligently explained to the Duma what he had been trying to accomplish in the interview. Sazonov assured the emperor that Protopopov had done no wrong and that the Ministry of Foreign Affairs was satisfied with his explanation. To his credit, the tsar was less convinced than was the press or the Duma.⁵⁵

Thus, by early September when the need for a new minister arose, Protopopov's name had been well recommended at court. Endorsements came from many varied sources; most of them had absolutely no connection with Rasputin, and indeed were usually hostile to him. Rasputin was only one of many speaking on his behalf, and this nod of assent came quite late, on the eve of the appointment, in fact. Plainly, he was not one of the man's most eager supporters. Badmaev, who had an exaggerated but typical estimation of Rasputin's worth, slanted part of his campaign for Protopopov in a direction calculated to please the *starets*. Around town the deputy revealed his conviction that he was divinely entrusted with the task of saving the country.

Rasputin was made aware of these facts in personal interviews with Badmaev. When Protopopov publicly proclaimed that he worshipped the emperor, Rasputin was somewhat impressed. Badmaev wrote letters to the tsar, Alexandra and Vyrubova, pointing out the virtues of his client. To Rasputin he sent assurances that the deputy had great affection for him and would do whatever he asked; he was a long-time friend of Kurlov; he swore he would have the police protect the *starets*; and he would release Sukhomlinov.⁵⁶ This was the kind of word Rasputin wanted to hear, but he had been too often disappointed in the past to have much

---

53. V.P. Semennikov, *Romanovy i germanskie vliiania, 1914-1917* (Leningrad, 1929), pp. 53-55.

54. K.D. Nabokoff, *The Ordeal of a Diplomat* (London: Duckworth and Co., 1921), p. 53.

55. N to A, July 20, 1916. Nicholas had just met Protopopov and said, "I saw a man whom I liked." But on September 9, he expressed reservations and wanted more time to think. On September 23, he spoke to Protopopov again and still was not certain. His mood was the same on September 29.

56. A to N, September 27, 1916.

faith in such assurances now.

On September 7, 1916 the empress wrote to Nicholas and endorsed Protopopov. Two days later the emperor revealed what kind of mood he was in and why Alexandra had moved so quickly to accept the candidate. At first Nicholas leaned heavily on her for emotional support, asking for her advice and consolation in the current difficulties. Then he reasserted himself and rejected the advice she had already given him. He complained that his head was spinning with confusion because of all the changes in ministers of late. He added that he was certain that this was bad for the government. As a footnote, he volunteered to say that he did not think too highly of the opinions of Rasputin and he was especially annoyed by the *starets'* latest recommendation, Protopopov. On the next day Alexandra repeated her advice, indicating that the deputy was the only one available for the post. Nicholas then made a brief trip to Tsarskoe Selo at the end of the month, and while he was there Protopopov was nominated Minister of Interior, but the emperor took the unusual step of naming him Acting Minister, a vote of scant confidence.

Protopopov found it easier to take office than to hold it. The Duma immediately denounced him.[57] The moderates had hoped to see one of their own men go into the council but they wanted him to enter in parliamentary fashion and to take office only after submitting his own set of conditions, which would be framed after consultation with the Duma. Protopopov had instead accepted the portfolio in the traditional way that made him a servant of the tsar. Immediately a campaign of vituperation was launched, accusing him of being not merely a creature of the hated regime but also a traitor to the Duma and his former friends.

Two principal charges were used against him: he had committed a treasonous act while he was in Stockholm, and he was an appointee of Rasputin. The first accusation could not be pressed too hard without making the Duma seem inconsistent, for it had accepted the minister's own explanation of what had happened in Sweden. But when the Duma met early in November both Miliukov and Purishkevich hammered at the theme that the government was betraying the country and that it was about to make peace with Germany. Consequently, for the sake of consistency, the charge of treason had to be used in the case of Protopopov, even though the Duma was repudiating its first reaction to Protopopov's explanation of his conduct in Stockholm. While there was some reluctance to adopt this tactic in dealing with the minister, there was no reluc-

---

57. P.N. Miliukov, *Istoriia vtoroi russkoi revoliutsii* (Sofia, 1921-23), I, 32.

tance to claim that he had been appointed by Rasputin and that he was going to carry out all of his pledges to the *starets*; these included an early peace with Germany. The raucous attacks in the Duma on Rasputin and Alix, along with references to treason, created fear at court.

Protopopov was shocked by the strength of the campaign against him; his reaction to it disappointed and angered Rasputin. The *starets* was concerned about the shortages of supplies. He could not understand why the large stores of food he saw in Siberia and the Caucasus could not be brought to the capital. He claimed that the officials who were responsible were busy talking and writing reports while people went hungry. There had been many things about Protopopov that Rasputin did not like, but the minister had agreed with much zeal that the time had come for vigorous action, and he had promised to help solve the food crisis by accepting and acting on some of Rasputin's nostrums. However, once in office he began to shrink from the responsibility he had formerly sworn to accept. He protested that his original idea—that only a man invested with dictatorial powers could manage the problem of supplies—was unworkable because he had proposed it when he expected the support of his Duma friends. Now he was under bitter attack from them. He was still close enough to several of the deputies to know of the scalding speeches that were in preparation for later dates. He believed that if he were to accept more responsibility in the government, the Duma, which was already clamoring for his dismissal, would be only further enraged. Therefore, he begged the empress to approach the emperor with the suggestion that the nomination to the post of comptroller of supplies be put off for several weeks. His mind once again grew cloudy and in the council he delivered incoherent talks.[58]

Nicholas thought that Protopopov himself might be useful in the position. The tsar had admitted that he had no notion of what was wrong with the supply situation and thinking about it made his head hurt, but he was sure that whatever the solution was it was a simple one. Only a few words and orders needed to be used to solve it. He was willing to permit Protopopov to occupy himself with this problem, so when Alexandra later sent Nicholas an outline of a *ukase* that would make Protopopov comptroller, the emperor agreed to issue it.[59] But once again, to the public it seemed that Nicholas was bowing to mere dictates of Alix and Rasputin.

---

58. *PTR*, Pokrovsky, V, 356-57.
59. N to A, October 31, 1916.

Rasputin was now at the height of his importance and he busily concerned himself with several affairs of state. For weeks he had been eager to see Protopopov given control of the supply situation. But when he discovered that the minister feared to undertake the task, in spite of previous commitments to it, he became furious and denounced him to Alexandra. She asked her husband to warn Protopopov to listen to Rasputin, a suggestion that was ignored although she repeated it many times. Against the wishes of Protopopov the ukase naming him comptroller was proclaimed. Rasputin expressed approval of this act, and the emperor uttered a prayer that it would solve the vexing problem of provisioning the cities.

Within a few days Rasputin was engaged with the empress in an attempt to keep Protopopov in the Ministry of Interior. The assault in the Duma started shortly before Nicholas was visited by several members of the imperial family and a number of persons in the government, all of whom pleaded with him to appoint ministers who would be acceptable to the public. They warned that unless this were done the country would explode in revolution. All of them said that a preliminary step might include the firing of Protopopov and the banishment of Rasputin. Some even suggested that the empress should be sent to a nunnery or to Livadia.[60] In a letter to Alix, Nicholas suddenly revealed he intended to fire Protopopov because he was mentally unbalanced.[61] Because Rasputin's fate seemed to be linked with the minister's, he joined the empress in attacking the suggestions of the Romanovs.[62] Alexandra sent Nicholas several pleas begging him to hold on to power a while longer.[63] She added that neither she nor the *starets* could at the moment think of anyone to replace Protopopov.

She was convinced that he was the last hope of the dynasty. He was the only member of the government who had any faith in the future. The others were being cautious now because there was a general expectation that soon they might be facing a new regime. Protopopov never admitted such a thing might happen. The empress was inclined at times to concede that it might be possible for a revolutionary situation to develop, but she believed a firm government could crush the threat. At moments when his mind became clouded Protopopov raved that he had been chosen by God to save Russia. He too admitted that there was some danger of revolution,

---

60. V.P. Semennikov (ed.), *Lettres des grand ducs à Nicolas II* (Paris, 1926), pp. 258-60.
61. N to A November 11, 1916.
62. A to N, November 4, 1916.
63. A to N, November 11, 12, 1916.

but he cried that if it came he would be ready to deal with it. He used the language that the empress approved of—he said that there would be hecatombs, rivers of blood would flow and then the revolution would be drowned.[64] He boasted that he had placed machine guns on roofs of houses so that mobs could be shot down. Although his enemies accepted this remark as the truth, the guns had really been placed in these spots in 1915 for air defense against German dirigibles.

Alexandra thought that Protopopov could be counted on to defend the government. She never admitted that he was demented; several times in reply to her husband she insisted that he was not mad. She claimed that he was a source of many ideas although he did not seem to be able to find ways to carry them out. There was a moment when she believed she had found a solution to this problem. She approached Mosolov and asked him if he would become a kind of administrative head of the government, using Protopopov's ideas which he would be charged to implement. She reminded Mosolov of his considerable experience. She did not understand when he protested that he had been a military officer and courtier all his life and knew little of public affairs or politics.[65] Government was to her not a matter of statecraft, it was a matter of being firm. Anyone who had served the tsar in one post could serve him in any other post. However, she did not reveal her plan to Nicholas. When it failed to develop she searched for some other way to retain Protopopov without having to suffer from his strange paralysis of will. Rasputin expressed approval of her efforts, but he was not able to make any suggestions as to how the minister might be retained in the face of the growing restlessness of the tsar.

Alexandra for the most part obeyed her husband's warning to keep Rasputin out of the dismissal of Protopopov, but she made a trip to the Stavka to convince Nicholas that the dismissal was a bad move since there was no one to replace him. Even the *starets* was unable to suggest a successor. Moreover, if Nicholas had to let Stürmer go at the same time it would be very difficult for him to resist the conditions of acceptance that the next ministerial candidates might choose to offer because it would be obvious that the government was desperate to fill empty posts. She

---

64. In one of his crazy moods he made remarks quoted in *New Times*: "I will stop at nothing...The first things I shall do is to send them [revolutionaries] from the capital by car loads. But I will strangle the revolution no matter what the cost may be." Alexander Petrunkevich, *et al.*, *The Russian Revolution* (Cambridge, Mass.: Harvard University Press, 1918).

65. Mosolov, p. 174.

assured him that the firing of Protopopov would open a hole in the dike of resistance to the demands of the Duma and of the military chiefs who wanted a new, liberal set of ministers. It would be dangerous to relent after recent, violent speeches in the Duma because the nation might think that the government was giving in. She summarized her position by indicating how important the minister was to her "Don't pull the sticks away upon which I have found it possible to rest."[66] Nicholas gave in and temporarily accepted her requests although he refused to honor Protopopov with a permanent status in the council, retaining him as Acting Minister of Interior.[67] His feeling was that as soon as a suitable replacement was found Protopopov would have to go. It was plain that it was this last reason that had done the most to make him change his mind again.

For the moment the minister was saved, partly because the empress had made his tenure a subject of her personal crusade. During the next ten days she avoided mentioning Rasputin's name in connection with keeping Protopopov. From November 26 to December 4, Nicholas spent a few days at home where he and Alexandra had a serious argument over Protopopov's remaining in the council.[68] It was not until 28 December, a few days before the death of Rasputin, that she reminded Nicholas that Protopopov was still serving only as Acting Minister.

During all these weeks Protopopov was close to Rasputin, visiting him often and being seen on the streets on several occasions in his company. In his cups one night at this time Rasputin boasted to Manasevich-Manuilov that he and the empress were the real rulers of Russia because the tsar had been too weak to rule alone. Manasevich-Manuilov, who was nervously living under the threat of a delayed trial for blackmail, spread the rodomontade far and wide and added to it the fact that he and Rasputin were very close friends.

§

Rasputin was accused not only of having ministers appointed; it was believed that he could also bring about their fall. There are two ministers who were thought to have been struck down by him: Kokovtsev and Sazonov.

---

66. A to N, November 12, 1916.
67. A to N, December 15, 1916.
68. N to A, December 4, 1916.

Kokovtsev was dismissed in January, 1914. Once he abandoned the coalition of Rasputin's foes in 1912, Kokovtsev took no further interest in the *starets* and never mentioned his name at court. This was in a way a concession, and Rasputin was grateful for the respite during the quiet years from 1912-1914. It is hard to say which of his foes played the leading role in destroying him.[69] One thing is evident—all of them were active. He estranged the Duma by trying to be the constitution-slayer. He made an audible remark saying that Russia fortunately did not have a constitution, and he also led a strike of the ministers who boycotted the Duma. These acts alienated all of the progressive deputies, but at the same time Kokovtsev had failed to win the support of anyone on the Right. He had angered conservatives by reducing government subsidies to the political campaigns they conducted; in addition, he had fought against personal grants of money to friends of the government and to reactionary journals.[70] Among the ministers he was disliked. Sukhomlinov hated him because of his caution in releasing funds to the military. In 1911 Kokovtsev had defeated the efforts of the clever Krivoshein to have the Peasant Bank transferred to the Ministry of Agriculture. Krivoshein was actively working to have the more pliable I.L. Goremykin made Chairman of the Council. Finally, Kokovtsev was also a foe of Maklakov and Witte—both busy intriguers—and the empress, who showed her dislike by snubbing him at a public reception.

All of these people who were determined to drive him out of office had little in common, and when they achieved their goal they went their separate ways. But from September, 1913 they watched with common gratification while Prince Vladimir Meschersky used his libelous publication, *Muscovite*, to wage a campaign of vilification against Kokovtsev. When the chairman served his purpose by successfully negotiating a French loan in November 1913, his days were numbered.

Nicholas most likely let him go because he had estranged so many persons in the government and outside it that he was finding it difficult to push through even minor measures in the Duma. Nicholas could see that Kokovtsev was a devoted monarchist and a highly capable financier although he was unimaginative as a political leader. He had the courage to be loyal; on several occasions he had proved this when he defied the Duma, an attitude few tsarist statesmen had the boldness to assume at that time. But his loyalty was not unlimited. While he would not avoid a

---

69. Kokovtsev, pp. 419, 434. He listed Shcheglovitov, Khvostov, Rukhlov, Bark, Krivoshein.

70. Ibid, p. 351.

clash with the Duma on certain issues of importance, he would not fight to defend the autocracy over the question of Rasputin. To the empress this was the ultimate test of a minister's loyalty. Kokovtsev had failed. In his biography Kokovtsev frankly admitted that he had forged the alliance of his own enemies and he freely identified them. When he pointed to the most dangerous foes he mentioned the empress and Prince Meshchersky. It was later, during the war, that Rasputin's name was added to the list of his foes, and then he was given the principal role in the affair.

The sudden dismissal of Sazonov in the summer of 1916 was said to be the work of Rasputin. The reason for the fall of the minister has never been made completely clear, but at least it is evident that Nicholas did not intend that he should go in disgrace. Sazonov expected the end; he had already prepared his assistant, A.A. Neratov, to be the real manager of affairs in the Foreign Ministry while Sturmer held nominal power.[71] The emperor accepted this situation. Of all the ministers, Sazonov seemed to be the one who had a charmed political life. His devotion to the monarchy was well known, but on several occasions he defied the tsar with impunity. It was he who had led the ministerial revolt of August-September 1915, an act which temporarily won him the hatred of the empress.[72] But within a few days she had a kind word to say about him. At the end of September 1915 she rekindled her hostility toward him, but a friend who was with her when she was informed of Sazonov's dismissal reported that she thought the moment chosen for his end was a most inopportune one.[73]

Nicholas, who went to unusual lengths to honor him, then appointed him ambassador to England, a post in many respects more important than the one he had held. It was obvious that during the final stages of the war and during the peace-making most of Russia's difficult diplomatic problems would center on its relations with England, and Nicholas most likely wanted his best diplomat at the court of St. James. Although the French ambassador was not unhappy to see Sazonov go, the British ambassador, Sir George Buchanan, tried to save Sazonov from the change of jobs.[74] However he may merely have convinced the emperor of the correctness of his decision to send Sazonov to London, because Buchanan

---

71. C.J. Smith, *The Russian Struggle for Power* (New York: Philosophical Library, 1956), p. 405. In March, 1913 groups interested in an aggressive Russian foreign policy in the Balkans demonstrated in the streets of the capital against Sazonov whom they blamed for the country's pacifist policy. Rodzianko joined in this opposition.

72. A to N, September 6, 7, 20, 1915. N to A, September 7, 1915.

73. Buxhoeveden, p. 223.

74. Raymond Poincaré, Au service de la France. Neuf années de souvenirs (Paris,

told the tsar that the British government had almost boundless faith in Sazonov and he was highly regarded by all British diplomats.

There may have been other reasons for Nicholas's actions. Probably he had conceded too much to the Poles during the first year of the war. He made these concessions conscious of what he was doing, and he issued the imperial statements not over his own name but over those of the Grand Duke Nicholas and Sazonov. The duke was gone. Sazonov could not be expected to betray his own agreements; therefore, he had to go. In the summer of 1916 discussions of the Polish situation were being carried on, and this may have been the reason the emperor chose this moment to appoint a new minister.

There is no indication that Rasputin had anything to do with the dismissal. His path never crossed that of Sazonov. The minister had expressed disapproval of the *starets* but he had participated in no plots against him. One recent student of tsarist foreign policy thinks that Rasputin destroyed Sazonov because he was too friendly to the Poles, a people the *starets* is supposed to have disliked. We cannot ascertain Rasputin's precise opinion of the Poles, but at times he talked of other minority peoples in the empire, the Tatars and Crimea, for instance. On the basis of these remarks it is probably correct to describe him as a federalist, although he probably never heard of the word. He felt that peoples who did not want to be in the empire should be permitted to depart in peace, perhaps his version of an idea current at the time: self-determination of peoples.[75] He had no quarrel with Sazonov over the Polish question, and there is no evidence that he sought the minister's downfall. More important may be the fact that he did not boast after the dismissal that he had caused it—a posture he often liked to assume. He was heard, in fact, to lament that Stürmer had taken over the post as Foreign Minister.

§

We have seen something of Rasputin's role in the appointments and dismissals of major ministers. But his power was reputed to extend through all ranks of the government, including some of the lesser offices. In these cases the action is not as well documented as it is with members

---

1926-33), I 371, 378. G. Buchanan, My Mission to Russia (Boston: Cassell Co., 1923), II, 187-88.

75. Simanovich, pp. 95, 107. "The blood of minorities is precious."

of the Council of Ministers; rumor and guesswork often play a more important part. A typical case involves the firing of the Assistant Minister of Interior, Vladimir Dzhunkovsky in September 1915, an incident which created a popular sensation.[76]

Dzhunkovsky had been a friend of the tsar from the time they had both been associated with the Preobrazhensky Regiment. Although he profited from his friendship, rising quickly through the military ranks to become a major general and governor of Moscow in 1905, Dzhunkovsky was regarded as an efficient administrator. With his vigor and honesty he created enemies, however he was well thought of in the Duma and he had the unwavering support and encouragement of the tsar.

Suddenly, in mid-August 1915 he was relieved of his post.[77] There was considerable speculation about what had caused this dismissal, but for the moment few facts were available. To complicate the picture, the general was given command of a guards regiment fighting in the Caucasus. Commentators were not certain that he had really suffered a disgrace, for it was generally believed that officers who served throughout the war in Tsarskoe Selo would have a hard time after the war justifying their bravery and patriotism, and consequently their careers would suffer. Dzhunkovsky's departure might be thought of as a promotion, or at least an opportunity to further his career on a front where Russian armies were winning battles. However, it was finally decided that he had been fired because he objected to Rasputin.

The accusers went back to a night in March 1915 when Rasputin had created a drunken scene in the Yar, a well-known restaurant in Moscow. For several weeks he had been claiming that a number of projects he had sponsored had been favorably accomplished. No one knew what these were, nor would he reveal anything. Most likely, this was one of his many attempts to build around himself the aura of a man who got things done. So important were these mysterious achievements, he said, that he was going to Moscow to offer prayers of thanksgiving before several of the national shrines. It was during this visit that he created the disturbance. An eyewitness, Bruce Lockhart, said that there was some noise coming from a private dining room—raised voices, chairs falling—but that was all.[78] The management was so awed by the *starets'* reputation that they handled him with care. The local police refused to arrest him. Eventually, a call was put through to St. Petersburg where Dzhunkovsky

---

76. *PTR*, Dzhunkovsky, V, 88-122.
77. A. Polivanov, *Memuary* (Moscow, 1924), p. 199.
78. R.H.B.

issued orders for the arrest.

Two versions of the incident circulated. One was in agreement with Lockhart, the other was much more lurid.[79] The head of the Moscow police, General Adrianov, gave the second version to Dzhunkovsky and Maklakov. In the course of a routine report, Maklakov presented to the tsar the milder account of what had happened. Dzhunkovsky then asked the police for more information on the affair. He consulted General Popov, former head of the St. Petersburg Okhrana. After the gossip had inflated the incident and added all sorts of imaginary details, Popov went to Moscow. He brought back to Dzhunkovsky an account that accused Rasputin of committing obscene acts and bragging that he controlled "mama and papa."[80] Dzhunkovsky told some of these incidents to the emperor, who listened carefully. Nicholas asked him to leave behind the written report. Later Dzhunkovsky privately circulated a copy of the document to friends in the court and in the Duma. The emperor, who probably did not read the written version, was under the impression that it contained only the material that had been originally given him in Dzhunkovsky's oral presentation; he wondered now why it created such a sensation. The empress, however, read it in June and denounced it as "that filthy paper."[81] She insisted that Nicholas "scream at Dzhunkovsky" and order him to destroy the report. Her tone was so harsh and scolding that she later felt compelled to apologize for her words.

Throughout this part of the incident Rasputin was meekly quiet and contrite.[82] He watched anxiously while the empress sent an aide to Moscow to ferret out facts that could be used to offset the police version of what had happened at the Yar. There was no indication that Nicholas was displeased with Dzhunkovsky, who during the summer accompanied him on several journeys. The incident of the Yar, which had been briefly discussed in March and June, was soon forgotten.

However, some people thought that the time had come to make another attempt to get rid of him. The moving spirit behind this attempt was the Grand Duchess Elizabeth, who sent one of her friends, a sister of Dzhunkovsky, to plead with the general to do something to rid the court of the *starets*.[83] Elizabeth added word that the Minister of Interior, Prince

---

79. PP Zavzarin, *Souvenirs d'un chef de l'okhrana, 1900-1917* (Paris, 1930), pp. 274-79. Paléologue, *Memoires*, I, 331-33.
80. Bienstock, p. 146.
81. A to N, June 22, 1915.
82. *PTR*, Beletsky, IV, 151-52.
83. Maurice Paléologue, *Aux portes du jugement dernier. Elizabeth-Féodorowna* (Paris,

Shcherbatov and the head of the Holy Synod, Samarin, as well as many influential people in Moscow, were supporting her plea. Dzhunkovsky was at first reluctant, but he finally submitted another report on the Yar episode to the emperor on 4 August. After reading it Nicholas grew angry at Rasputin, and on 5 August ordered him to go to stay in Siberia. It was six weeks before he was able to return to the capital; this was accomplished only after the most intensive campaign by the empress on his behalf.

On 6 August, Minister of War Polivanov broke an oath of secrecy and stunned the council with word that the emperor would take command of the armies. The emperor was on the defensive, and the opponents of Rasputin thought this might be a good time to make their victory complete. They sought permanent banishment for the *starets*. A partial victory made them careless. On 15 August an influential financial journal attacked Rasputin, and asked how he had got his hands on all of the important branches of government. The article extensively quoted indirectly many passages from Dzhunkovsky's report. This was a serious offense against court protocol because reports given to the tsar could be made public only with his permission. Prince Vladimir Orlov, a friend of the emperor and one of the highest ranking courtiers, made a statement that it would be good for the country and the regime if Alexandra were sent to a convent or to Livadia for the remainder of the war.[84] Dzhunkovsky was indiscreet enough to join Orlov in making this recommendation. At the same time Sazonov hinted to the emperor that the suggestion might have been a wise one. The public believed that the empress might soon be leaving for the Crimea.

She immediately launched her counterattack, demanding that both Orlov and Dzhunkovsky be fired. She noted that the latter was a friend of Guchkov, and that he had once deliberately failed to prevent Iliodor from smuggling abroad a scandalous book that contained many lies about her. She regarded him as a personal foe because of his approval of the idea that she should be exiled. She hardly mentioned that he was an enemy of Rasputin; the empress was fighting for herself and not for her friend.

The attacks had gone too far. Nicholas had been most disturbed by the publication of Dzhunkovsky's report; he was generally chagrined that an old friend permitted this breach of court discipline. At the same time he was becoming more dissatisfied with Shcherbatov because the prince could not or would not muzzle the press when it lamented the change of

---

1940), pp. 139-144.

84. *PTR*, Dubensky, VI, 386.

command. The two offices, Minister and Assistant Minister of Interior, were regarded as almost one. The minister was usually permitted a voice in the nomination of the assistant. On 15 August, Nicholas, following a meeting of the Council of Ministers, told Shcherbatov that he was thinking of nominating Beletsky to the post of Assistant Minister. This was a threat to the prince because he and Beletsky were enemies. Such a nomination could mean only that the minister was on shaky ground. Nicholas was vexed at Dzhunkovsky and now he found an additional reason for wanting him to leave his post.

Dzhunkovsky had become a liability to the emperor. On 16 August he was relieved of his duties at court, and in a short time he was ordered to assume a command in the Caucasus. The empress regarded the move as a promotion; she complained to her husband that this was no punishment, and in her letters for the next few months she renewed her warnings about the general.[85] Nicholas warned her that such things were none of her business.

The public believed that Dzhunkovsky fell because he had clashed with Rasputin and often feuded with him since March 1915.[86] The Yar affair was now resurrected and regarded as the cause of the fight between the two. Few knew or cared about the real reasons for the dismissal. Only the Grand Duke Andrei Vladimirovich correctly guessed that the publication of parts of the report was one of the reasons for the general's downfall. Alexandra called the attention of her husband to the rumors which attributed the dismissal to a fight with Rasputin. It was obvious that she knew that this was not the reason for the dismissal.

§

Cries of anguish came from the empress and from Rasputin when it was announced in July, 1916 that Alexander Makarov was to become Minister of Justice. Makarov had been unsuccessful in silencing the press at the height of the controversy over Hermogen in 1911 and 1912; it had been his contention that the government under the law had the right only to request that the subject not be mentioned. He believed that any attempt to muzzle the press was illegal. Although Nicholas did not like this

---

85. A to N, October 6, 8, 1915.
86. Paléologue, *Memoires*, II, 72.

reply, he accepted it. Makarov had eventually succeeded in obtaining the compromising letters of Alexandra from Iliodor, but instead of quietly placing them in the hands of the empress he impulsively gave them to the emperor. Several of Makarov's friends predicted that this would bring about his downfall.[87] But he remained in office for eight months. In April 1912 he had offended the Duma in the debates over the Lena Massacre when he said, with a shrug of the shoulders, that incidents such as this would always happen from time to time because it was the way of the world. When he was finally let go there was no campaign in progress to remove him. While he may have offended Rasputin in 1911 for his sluggishness in dealing with the press this was not the thing that caused his end. Politically he had gotten a bad reputation both at court and in the Duma; this made his dismissal inevitable because Nicholas probably did not care to suffer the burden of holding on to a minister who was disliked by the Duma and the court.

During the next few years Makarov was recognized as a foe of Rasputin.[88] This did not prevent the tsar from considering him for a cabinet post. We do not know who pushed this appointment but it had the marks of haste and desperation. In May, 1916 Rasputin alerted the empress to rumors in the capital that Makarov might once again be picked to succeed the elder Khvostov as Minister of Justice. She wrote immediately to remind her husband that the man hated Rasputin and would not protect him, nor would he protect her by keeping her name out of the papers.[89] Nicholas warned her that he did not want to be interfered with in such matters.[90] When Makarov was finally appointed two months later, she reminded her husband once more that ministers who did not like the *starets* brought bad luck, and she asked that special measures be taken to protect Rasputin.[91] Nicholas did not reply. In office, Makarov continued to be anti-Rasputin and forced the prosecution of Manasevich-Manuilov.[92]

Rasputin was interested in the Ministry of War. From time to time he requested the empress to pass on his advice in military matters to the emperor. He also asked that Nicholas order his generals not to attack.

---

87. *PTR*, Beletsky, IV, 149. V.I. Gurko, *Features and Figures of the Past* (Stanford, Stanford University Press, 1939), p. 52.
88. A.V. Guérassimov (Guerasimov), *Tsarisme et terrorisme* (Paris, 1934), p. 294.
89. A to N, May 23, 1915.
90. N to A, May 24, 1915. This letter also contained a threat aimed at Rasputin.
91. A to N, July 16, 1915.
92. *PTR*, Makarov, II, 128.

Rasputin was obsessed with the idea that the large number of casualties in the armies were undermining the regime, and he wanted to see the army play a passive role in the war because the highest losses came during attacks. In mid-June, 1915 neither Rasputin nor the empress approved of the appointment of General A.A. Polivanov, who was determined to provision the armies for an offensive war. The empress did not like him because he was a friend of Guchkov and had openly courted approval of the Duma. Nicholas had warned him about his associations, but he did not take heed of his wife's pleas to keep the general out of the cabinet.[93] At this time Nicholas made a series of ministerial appointments displeasing to Alix. When Polivanov was nominated Nicholas informed her that he was consulting Alexander Krivoshein, whom she did not trust, about other nominations, but he did not tell her what they were nor did he invite her opinions.

Polivanov cooperated closely with the special councils, the Duma and the opposition leaders. As a result he was fired in June 1916. At the Stavka friends warned him of his impending dismissal; his close ties with the Duma had made this a certainty. In other words, he had done many things to displease the emperor. But when he fell from power he blamed Rasputin for his fate saying that the incident in which he had taken a military car from the *starets* had caused his end.

Alexandra wanted a more pliable Minister of War to follow Polivanov. Rasputin joined her in this wish and suggested General N.I. Ivanov.[94] This was a rare instance of Alexandra admitting she and Rasputin disagreed, for her candidate was General M.A. Beliaev, Assistant Minister of War. Nicholas later remarked that Beliaev was a weakling and unable to get things done.[95] Without waiting to hear her recommendation, Nicholas announced that he was appointing General D.S. Shuvaev, an officer at the Stavka.[96] Alexandra had heard rumors of this appointment and had specifically warned her husband that this man would not be able to deal with the Duma. In the summer of 1916 she began to attack Shuvaev and continued to promote the cause of Beliaev.[97] Nicholas responded by firing Beliaev as Assistant Minister of War. This ended her interference for a long time, but in December 1916 she once again reminded her hus-

---

93. N to A, January 12, 1915.
94. A to N, March 13, 14, 1916.
95. N to A, August 16, 1916.
96. Shuvaev had not sought the job and was surprised when he was named. *PTR*, Shuvaev, VII, 285.
97. A to N, August 13, 1916.

band that Beliaev would make a better minister than Shuvaev. Finally, a few weeks before the revolution, a desperate Nicholas gave in and made Beliaev Minister of War. Rasputin stayed clear of this bickering, and once the successor to Polivanov had been named, he took no part in the arguing over the post. However, he continued to make an occasional request about the need to avoid losses of men.

As a *starets* at court Rasputin felt he had a right to exercise his prerogative to intervene in certain issues that had nothing to do with appointments. These concerned his ministry, and therefore he spoke with assurance and without apology. Usually he did not pass on this kind of advice through Vyrubova; he boldly brought it to the empress who was supposed to bring it to her husband. He had no hesitation or caution in these matters; he vigorously stated what he believed and sometimes added a warning that if he did not have his way then evil would befall the family or the nation. His actions in such affairs were quite different from his behavior in the cases that are more clearly political. It is obvious that Rasputin was plainly aware of the difference in the two kinds of intervention he was indulging in, and he adjusted his tactics accordingly. But even when he spoke as a Man of God his degree of success was not great.

In June 1915 after the start of the great defeats on the Eastern Front, the Minister of War, V.A. Sukhomlinov was fired, sacrificed because of his ineptitude and because he was a victim in a behind-the-scenes political struggle that had been going on for many years. A.A. Guchkov since the spring of 1908 had been engaged in a long-range attempt to wrest control of the army from the tsar, who by law was in charge of all military affairs. Without being a revolutionist and without associating with revolutionaries, Guchkov, a political moderate and a resourceful and ambitious schemer, most likely planned to use the army someday in the struggle for power between the tsar and Duma. In this deadly game the stakes were high—especially for the loser. In March, 1915 his old foe Col. S.N. Miasoedov was accused of spying. A kind of drumhead courtmartial tried, convicted, and shot him all in one day. Although Miasoedov had a shady past, he was almost certainly not guilty of spying, and the trial was a mockery of justice. But Guchkov had showed his power in military circles.[98]

For the tsar to retain Sukhomlinov in office after the start of the outcry caused by the defeats would have given Guchkov and the Duma

---

98. For a realistic appraisal of the politics of Guchkov see George Katkov, *Russia 1917. The February Revolution* (New York: Harper & Row, Publishers, 1967), pp. 129 *passim*.

an issue that could be used effectively against the government. Guchkov would have welcomed the chance to pillory that general, for the pair had been enemies since 1909 when the general had been made Minister of War with the assignment of stopping Guchkov's drive for influence in the army. In addition the general was too good a target; the daughter of the British ambassador described him as "The true picture of the drawing room soldier, scented and pomaded, with gold chain bracelets on his white wrists, an unctuous manner and a pretty wife whose antecedents were unknown."[99]

Although his removal was justified by his failures and blunders, he was disappointed that the tsar did not stand by him. But public opinion had been heated by another widespread campaign of gossip apparently emanating from the Duma and vilifying the regime with hints that some of its people indulged in treason. Popular fear had begun to demand proof for its suspicions and satisfaction of its lust for scapegoats. The emperor liked the general and thought he was negligent rather than criminal,[100] but to deflect criticism from the regime in the spring of 1916 he permitted him to be deprived of the privilege of house arrest and to be put in the Peter and Paul Fortress. He expressed to the Minister of Justice, the older Khvostov, his displeasure at the proceedings, calling them "disgusting."

Rasputin did not step into this whirlpool of intrigue because he was primarily interested in its political aspects. In fact, political considerations alone would have kept him out since the situation was loaded with danger for him. Once it became known that he was meddling in the matter, the gossip and the newspapers turned on him. He had not tried to save the minister from dismissal and he did not become involved until nine months later when Sukhomlinov's wife gave him many facts, reporting that according to doctors the general was beginning to lose his health and sanity in prison. There were also rumors that he would later suffer the final immolation: he was to be shot. Rasputin could feel genuine compassion for victims of official mistakes and miscalculations. If he was convinced that Sukhomlinov was also involved in minor embezzlements that would make little difference. He believed that trials, punishment and retribution all belonged to the Lord. Thus in the name of humanitarian considerations he thought it was his duty to speak out and to direct attention to an injustice.

Alexandra had at first taken little interest in Sukhomlinov's fate; she strongly disliked his young and glamorous wife and was inclined to

---

99. M. Buchanan, *Dissolution of an Empire* (London: J. Murray, 1932), p. 107.
100. Polovanov, p. 141. N to A, April 30, 1916.

think he had gotten his just deserts. She even compared him to the new Minister of War, Polivanov, whom she suspected of treason.[101] When Rasputin began to intervene with humanitarian arguments on the general's behalf, she did not bend much but only indicated to her husband laconically that imprisonment might not be good for Sukhomlinov. But when her Friend of God began to use political arguments she listened more attentively. At the same time Badmaev had entered the case, appealing to the tsar for clemency.[102] Nicholas may have expressed displeasure with these intrusions because for the next few months when she mentioned the case she was cautious and apologetic, indicating clearly that she was being pressed by Rasputin. When she passed on letters from Madame Sukhomlinov she asked Nicholas's indulgence for her intrusion. Next she nervously informed him that she had warned Vyrubova and Rasputin that she thought her husband would not want to mix in the affair anymore; meanwhile, she passed on from Rasputin information that the trial could not possibly be honest because the judge had once been fired from a military post by the accused.[103] A week later Rasputin had another inspiration: he thought that Sukhomlinov might be released quietly after the government released news of the next victory (The successful Russian offensive of the summer of 1916 was about to get underway).[104] Rasputin grew desperate; his reputation was at stake. If he could not induce the tsar to perform this simple act of mercy the picture of him as a man of influence might evaporate. His importunings soon made the empress use stronger words in asking that the general be helped. But Nicholas stood firm for a long time. In the autumn the crescendo of her pleas increased.[105] Finally, Nicholas gave in, but he only released Sukhomlinov to go to his own home. The threat of a trial still hung over him. On the next day she replied to Nicholas's slow generosity with a curt "thanks," placed near the end of a letter.

In the summer of 1915 Rasputin began to worry about the lack of fuel and food in the cities.[106] Alix was not as convinced as her friend was of the gravity of the question but she passed on his words nevertheless. His concern was eventually reflected in her letters which in September

---

101. A to N, March 4, 1916.
102. Semennikov, *Za kulisami*, p. 25.
103. A to N, June 9, 1916.
104. A to N, June 14, 1916.
105. A to N, September 22, 27, 1915; November 1, 9, 1916.
106. *PTR*, Goremykin, II, 17. A summary of this problem and some of its causes can be found in P.B. Struve, *Food Supply in Russia During the World War* (New Haven: Yale University Press, 1930), pp. xiv-xx.

began to call attention to the problem. Later, when supplies became a matter of appointments, she grew more interested. Rasputin at first had little confidence in the notion that provisioning could be helped by putting the right man in charge. He thought, on the other hand, that a few orders by the emperor could solve all difficulties. He wanted the emperor to send inspectors to plants, factories, and rail yards where their presence would have a magic effect. He tried to bring pressure to bear on the emperor from several directions, requesting Goremykin to personally bring the matter to the attention of Nicholas, and at the same time asking Alexandra to mention it in her letters.[107] He drove home his point in a long discussion with her during an entire two hour interview in October, 1915. His plan, which was eventually carried out, involved the suspension of all westbound rail traffic for three days while nothing but food trains went along the track. But the project failed because the authorities neglected to gather in advance the necessary shipments at the railheads. Confusion brought on a rash of speculation and hoarding. Storekeepers unexpectedly ran out of goods, and then suddenly without explanation they had oversupply. All of this hurt morale in the cities.

But Rasputin was convinced that it was the shortages of food and fuel that would bring ruin to the country; he reported to the empress that he had seen this in a vision.

It was through the entanglements of this problem that he came by an interest in the cause of Protopopov. Alexandra reported that the ministers were discussing the need of a plan.[108] Rasputin did not like talks and conferences. If someone could be found who would issue simple and uncomplicated orders from the emperor the matter might be solved, he reasoned. Urgency demanded immediate action and not further study; hungry and cold people in the cities were thinking dark thoughts and contemplating violent deeds. His new scheme called for a strong minister to serve as dictator of supply problems. In the meanwhile the subject caused controversy between several branches of the government, the Ministry of Ways of Communication and the Ministry of Agriculture arguing over jurisdiction, one with the other and each with the Duma.[109]

By June 1916 he had come to the conclusion that the Ministry of Agriculture ought to step out of the picture. With large stores of food available in remote sections of the country there was no need to increase production, the usual concern of this department. It was a question of

---

107. A to N, October 10, 1915.
108. A to N, December 19, 1915.
109. A to N, January 1, 1916.

getting available food to the consumers. He thought that the Ministry of Interior could best solve this problem because it was equipped with police powers. He then moved into the realm of appointments and dismissals in connection with this question. Although normally he was well disposed to a man who treated him with any measure of deference, he said that he believed it would be best if the genial Governor General of St. Petersburg, Prince A.N. Obolensky, were dismissed because he was incapable of getting food and fuel to the city.[110] The emperor replied that he found all of this talk boring and thought that other things were more important. But Rasputin did not give up; he felt that this was a province that was his because he was talking not of logistics but of human suffering. By the autumn of 1916 Nicholas hoped to solve the matter by appointment of his own nominee; unfortunately he could find no one and was therefore forced to accept Protopopov. Shortly before his death, Rasputin repeated that the problem of provisioning the cities was the most critical problem the regime had to deal with.

When the government announced in June, 1915 that it would soon call up the so-called second category of reserves, Rasputin expressed very strong opposition to the plan.[111] The men in this class were between twenty-one and forty-three years of age and were untrained. Only twice before had they been called up: during the Napoleonic invasion and during the Crimean War. Aside from the fact that the call admitted that the country was in the midst of a desperate military crisis, there were other things wrong with it so far as Rasputin was concerned. He knew that villages during the sowing and harvesting seasons were beginning to feel the pinch of the manpower shortage. Most families thought that they could rely on the men of the second category to stay with them. Now, before the harvest, they were going to suffer the shock of losing them. Rasputin spoke on their behalf and protested strongly against the plan.[112] The empress was not in a good position to take up his cause. She had just failed to convince her husband of the truth of her curious belief that General Ivanov was a spy, an idea that the emperor had termed "not worth an empty eggshell."[113] Still, we can feel the pulse of Rasputin's fears in the sudden and strongly worded pleas that she forwarded. She said that Rasputin begged that for the sake of internal peace the men be left in the villages. St. Petersburg was already crowded with several hundred

---

110. A to N, February 1, 1916.
111. A to N, January 10, 1915.
112. A to N, June 23, 24, 1915.
113. N to A, September 9, 1915; June 14, 1916.

thousand peasant-soldiers who had only recently been inducted into the army and who were crowded into barracks without any officers to lead or discipline them. Revolutionary agitators were working busily among them. Adding many more thousands of recruits to this untrained mob would only increase the threat to public order in the capital. Rasputin predicted eventual tragedy if the second category was called.

After three weeks of hesitation Nicholas deferred his order to call the reserves.[114] This was not the same thing as an outright cancellation. We have no way of knowing how much weight the words of Rasputin carried.[115] But the unusual force he used to express his views may have had some effect.

A similar problem existed with the deportees from the areas of war. In August, 1915, following the great retreats of the previous months, crowds of refugees from the West were camped around the capital; they formed a vast horde of people living in misery. Many were Jews, Poles, and Lithuanians who had little interest in the war and were spreading a spirit of defeatism near the citadel of the Russian government. They had been brutally forced by military authorities to leave their homes, often on a few hours notice, and to travel for hundreds of miles. They arrived in the hinterland starving, with no possessions, and having lost many of their number in death. Huddled in squalid shanties or packed into tenements inside the capital, they formed a combustible mass. Rasputin regarded them as a danger to the regime since they did much to sow a feeling of apathy and hopeless weariness.[116] He repeatedly warned the empress about the situation and on one occasion he induced her to go on a personal tour of some of the camps. Although she tried to impress her husband with Rasputin's conviction of the necessity for action, neither she nor the emperor appreciated the importance of the problem, and her words therefore lacked urgency. Nicholas dismissed her admonition by retorting, incorrectly, that the refugees had fled their homes at their own will.[117] After a short time the subject disappeared from the empress's letters.

Rasputin was much concerned with trivial matters which were of no interest to the government. He realized that in these affairs, however, the little man's estimate of the regime was often formed. Rasputin was always on the lookout for such situations because he knew that the of-

---

114. A to N, June 29, 1916.
115. "Rasputin...'Okhranki'," *KA*, 275.
116. A to N, August 27, 29, September 16, 1915.
117. N to A, August 27, 1915.

ficials were either indifferent or unaware of the problem that confronted the ordinary citizen. Thus he sought to have Nicholas rescind the order forbidding common soldiers from riding on the street cars in St. Petersburg. He also asked that the fares on these conveyances not be increased because this would work a hardship on the workers in the factories. In both cases Nicholas promptly gave his approval to the ideas of the *starets*. He even went so far as to express his gratitude for the alertness of Rasputin in discovering these abuses. In the estimate of Nicholas, Rasputin had found his rightful role. Encouraged, Rasputin, shortly before his death, asked that the wartime stamps which were officially recognized as currency be discontinued. He warned that the illiterate worker and peasant were being victimized by unscrupulous business people who underpaid them with the unfamiliar currency. Cab drivers in the city complained that in the dark of night some of their passengers were cheating by paying the fare in small denomination stamps which were the same dimension as stamps of larger value.

In such small matters Rasputin seemed to find his forte since both the emperor and the empress listened to him, but these are not the larger affairs like the appointment of ministers in which he was supposed to be playing an important role.

Perhaps the most poignant of all Rasputin's failures, and the one that speaks most eloquently about the extent of his real power, was his inability to keep his mentally abnormal son out of the army. The boy, Dmitri, was called up in mid-September 1915. Alexandra immediately requested her husband to do something about this although she made no specific recommendations.[118] She added that Rasputin was in despair. He could approach no one in the Ministry of War; he was hated there and had few friends in any part of the army. Having given up hope of keeping Dmitri at home he began to plead for a special assignment for him. From Pokrovskoe he sent a barrage of telegrams to the tsar asking that the boy be stationed at Tsarskoe Selo. Then he reduced his demand and asked that he be sent to a non-combatant medical regiment. At court Rasputin acidly remarked that only one Romanov had served briefly as a soldier in the war against Japan, and during the dark days of 1915-16 two of the young and healthy members of the family were notorious for their nightly escapades in the city and for their unwillingness to serve at the front. Ten days after the original request of Alexandra, Nicholas replied that it would be impossible to make any special arrangements for Dmitri Ras-

---

118. A to N, August 30, September 1, 11, 1915.

putin.[119] Vyrubova consulted V.N. Voekov, Commandant of the Palace, in an unsuccessful attempt to have the boy assigned to a special organization. Rasputin's son entered the Tiumen Regiment as an ordinary recruit.

§

The dark figure of Rasputin moved through the process of several ministerial appointments. The public suspected that he was always in the foreground because it thought that in each case he had interests at stake. But in not one of the cases examined did he make a minister, nor was his approval the principal ingredient that determined whether the appointment would be accomplished.

But he was really one of several participants in each plot, sharing importance with intermediaries such as Badmaev and Andronnikov, who might be insiders to bureaucratic intrigue but were outsiders at court, where they thought power resided and from which place it might be manipulated. These intermediaries played part of the role that the public thought Rasputin played almost alone. It was they who conceived the appointment, found the candidate, seized the initiative in the plot and defined its outlines, assigned Rasputin his role, and at all times kept the momentum of the plot moving forward.

With the exception of Trepov each candidate was active on his own behalf. He sought to send word to the tsar about his availability, and he saw to it that friends sent word through other channels. It was necessary to create the impression that a number of independent sources were clamoring for the appointment. Most aspirants believed that sponsorship at court was necessary. This was often difficult to achieve because the tsar and his wife were inaccessible. Moreover, too forward a policy by the potential minister would be unseemly. At this point some kind of intermediary would be most useful.

Enter Rasputin. He was contacted either by the candidate or his sponsor, who had access to sources of information that Rasputin could not always tap. He was supposed to find out what the tsar was thinking. This was no easy task since Alexandra, his ultimate informant was often unaware of her husband's inmost thoughts. In retrospect, we are struck by the fact that Rasputin at no time possessed esoteric infor-

---

119. N to A, September 9, 1915.

mation. For the most part he depended on his shrewdness in guessing what was going on. When he knew about a nomination pending, so did many others, including hangers-on like Andronnikov. On the other hand, Rasputin was often ignorant of the trend of events. Many of his trips to Siberia were timed so he would be out of town not only when he was under attack but also when he was being pressed to assist in an appointment that he really knew little about. In the critical days of the summer of 1915 when many cabinet changes were either made or rumored, and when a truly influential person would have wanted to be close to the capital, Rasputin was absent. He returned briefly from Siberia in late July and then after a few days quickly departed on his travels once more. When he verified the truth of the rumor of an appointment he still could not act upon it until he had some indication that the emperor was ready to make a move. If he committed himself prematurely, if he made a promise he could not keep, or if he predicted what did not come true he could be ruined. His supposed recommendation, or *demand*, as he called it, had to be carried out before he could be identified with it.

At all times it was essential for him to protect himself, to prepare for possible future excuses should the candidacy fail. As the political crisis of tsarism deepened this became less likely since there were almost no willing candidates. Those who were willing were going to Rasputin, among others, because of his reputation, and they were asking for his assistance. Thus through the operation of a kind of system, based on a false conception of his role, he was able to find out who was interested in serving the government and whose name he could mention at court. He became a kind of clearing house. He was making almost no choice among interested persons because there were no competing candidates.

The appointment of Khostov was preceded by the most elaborate dealings that Rasputin ever became involved in. At this time the empress and emperor both wanted to see the dismissal of Prince Sherbatov, whose presence in the cabinet was causing Alix to panic. Rasputin was never convinced that his interests and those of Khvostov were identical, therefore, although he reluctantly agreed with Alexandra, he did not share her sense of urgency, nor did he ever take the trouble to do his best to represent the aspiring minister at court. In his negative response to Khvostov's drive for power, Rasputin waited almost too long, and the candidate came near to taking office without his help. It can almost be said that Khvostov got the post in spite of Rasputin. To win it Khvostov plotted and campaigned with a diligence and concentrated effort that no other office-seeker had mustered. He thought he needed the help of Rasputin in two ways: the *starets* could give him some idea of what to say at court, and he could arrange the first interview with the empress, considered by Khvostov to be the key to power.

Khvostov finally fell not because Rasputin pulled him down but because he reached for more power and in reaching found he had to destroy Rasputin, a feat which he lacked the skill and courage to accomplish.

The plot to install Stürmer in office was already well underway when Rasputin was introduced into it. Through several independent avenues of approach, Stürmer had brought word to the tsar of his availability. He was a well-tried paladin of autocracy, a tarnished knight who had put his sword at the service of the emperor in a number of battles early in the reign. When he sought the ministry he had no competitors. All these things were in his favor, and they had nothing to do with Rasputin or his opinions. The *starets*, however, saw that a few of his own interests might be at stake. Although he suspected that Stürmer might be neither loyal to the emperor nor firm with the Duma, he feared his well-known cunning. Rasputin's lukewarm support might temporarily put fetters on the nominee. In addition, Stürmer was a possible counterweight to Khvostov, who was growing dangerous in his ambitions. And so Rasputin's misgivings were stilled.

When he heard that Trepov was to be named, he entered the lists, but the emperor had already made up his mind. Rasputin therefore moved sluggishly around the edge of the appointment process and could not even use the minister's one blunder, the offer of a bribe, to influence the course of events. The press and Duma deputies hated Trepov, thinking he was a creature of the *starets*. In fact the appointment signalled the end of Rasputin, for it encouraged the plotters, who had already prepared a plan to kill him. They knew, unlike the public and the deputies, that Trepov was not a friend but a foe of Rasputin, and as long as he was in office he could be counted on to discourage the officials from pursuing the murderers with too much vigor.

In the case of Protopopov, Rasputin had to move with care. It was not his support or that of the empress that won the day; once again it was the need of the emperor to appoint the only available man. Rasputin's powers of persuasion or his influence had little to do with it. But Protopopov, unlike other recent ministers, had wide support for his nomination, from the king of England to Paul Miliukov, at that time the leading spokesman of the Duma. Rodzianko, who was titular leader of groups normally sympathetic to the Duma and at the same time passionately devoted to monarchy, had stood firmly behind Protopopov. Nicholas was the only one who at first questioned the nominee's sanity. But Protopopov seemed to be loyal to the tsar and to have the support of the public and the Duma; it was hard for Nicholas to resist such a combination.

Both the empress and Rasputin fought to save Protopopov because Rasputin's very life depended on their success. For the same reason Rasputin opposed Trepov because he knew Trepov's agenda had the removal of Protopopov at the top of the list. If Protopopov remained in office Rasputin thought he could be assured of some personal protection, and at the same time he hoped to use the minister as a mouthpiece for his own ideas about the problem of provisioning the cities. However, the fact that Protopopov was unable to supply protection indicates his uselessness. Rasputin was killed while he was Minister of Interior, probably because of deliberately lax police protection. In addition, he did not have the courage, or sanity, to push through Rasputin's ideas about the food supply.

IX

WHERE THE ROAD ENDED

Rasputin's death, the last sensational incident of his career, was a minor happening played out against the backdrop of a great drama—the all of the Russian Empire. People were swept up in the movement and carried along to the final tragedy. With the end approaching, strong passions and panics gripped them. In Petrograd late in August 1915, as gloom settled on the country because of the great retreats of the army, silver coins disappeared from circulation. No one could find a rational cause behind the hoardings. Suddenly, without apparent reason, the coins appeared again.[1] One detached observer described the scene in this way:

> The forces of internal disintegration continued their destructive work, as if Russia had been doomed beyond redemption. The most incredible, sinister rumors spread everywhere. You could not enter a single home without hearing fantastic stories about happenings at court. To accept half of them would be to believe that people were going insane. As officers on leave were in most homes together with their orderlies, the gossip and ru-

---

1. G.T. Marye, *Nearing the End in Imperial Russia* (Philadephia: Dorrance & Co., 1928), p. 231.

mors spread through all classes of society...You found yourself repeating the same stories.²

The air of fantasy and expectancy increased. As the reckoning came near, people began to talk at an increased tempo, as if to drown their fears. In the midst of all this talk the name of Rasputin appeared more and more frequently. One young grand duchess recalled that

> All attention was riveted on interior events. Rasputin, Rasputin, Rasputin—it was like a refrain; his mistakes, his shocking personal conduct, his mysterious power. This power was tremendous; it was like dusk, enveloping all our world, eclipsing the sun. How could so pitiful a wretch throw so vast a shadow? It was inexplicable, maddening, baffling, almost incredible.³

In such an atmosphere, and with such a fantastic figure—Rasputin—it was possible to believe anything; therefore, stories of real and artificial scandals were mixed together and swirled around the throne. By this time his name was so closely linked with those of the royal couple that what was said of him could be said of them, and they came to share some of his evil reputation. With Rasputin the monarchy was shackled to a corpse.

He was the target of familiar charges that had been used against him in the past: heresy, corruption, and debauchery. During the war a new and dangerous accusation was added: treason. Most people who retained faith in the country and had some measure of confidence in the monarchy believed that the nation's defeats could be attributed to betrayals committed by individuals in the government or the court. Rasputin and the empress were thought to be the ring-leaders of a so-called German party which was gathering information and sending it to Berlin. The evidence to support this claim was always vague. For instance, at one time Rasputin was much interested in knowing if a Russian attack was about to take place on a certain part of the front. His inquiry was seized upon as proof that he was a spy. Later, however, it was learned that he was trying to help speculators who wanted to buy cheap timber rights in a forest held by the Germans. A successful attack might place the timber in Russian hands—with a consequent steep rise in value and a good

---

2. Dmitri I. Abriskossow, *Revelations of a Russian Diplomat* (Seattle: University of Washington Press, 1964), pp. 226-27.

3. Grand Duchess Marie, *Education of a Princess* (New York: Viking Press, 1935), pp. 248-49.

profit for the speculators. As usual, what people thought was Rasputin's sinister purpose was really the substance of one of his trivial interests. However, before the refutation had caught up with the charge, the public had moved on to more sensational things.

Although no one has ever offered convincing proof that he cooperated with German agents, the charge was widely believed during the war. Some people found it easy to take the next step and to think that the throne was involved in the betrayals. Therefore, of all the charges hurled at the *starets*, this one carried with it the most dangerous threat for the monarchy, and in the long run played an important part in dooming it. Many believed the accusation, at least in part, and some did not, but most were alarmed and uneasy about what they heard was going on at court. Gradually, they found their affection for the monarchy of Nicholas II disappearing. The monarchists of the upper classes, who would have been normally unshakeable in their support no matter what their complaints, became neutral or hostile. The alienation of the bureaucracy, nobility, and church, described earlier, now worsened. Rumors of a palace coup spread everywhere. The public showed a sudden interest in the fate of Tsar Paul, in 1801 murdered by officers who came to think of him as a menace to the nobility and the country.[4]

Monarchists had been educated to hold a deep respect for the throne. They looked to the autocracy with an honest reverence and deep emotional attachments that are hard for outsiders to understand. They referred to it as a thing of beauty, "a perfect rose" as Iliodor called it, and felt almost an aesthetic pleasure in contemplating it. But the presence of Rasputin ruined much of this. By 1916 many of them believed that Rasputin, "the filthy muzhik," as they referred to him, was befouling the throne. With him monarchy was becoming repugnant, a thing always near death but refusing to die. Its character was typified by shoddy deals, backstairs intrigue, and corruption. Late in September 1916, the saddened Guchkov, holding out his head in disbelief at what was happening and thinking of Rasputin and Nicholas, said to a friend:

> Here is this great empire, with all the tradition of the Great Peter, shrivelled into an insignificant little man with no will in him. Here is the great Orthodox Church with its gorgeous ceremonial and music, represented by a filthy, snuffy lay-brother, a charlatan and adventurer.[5]

---

4. B. Maklakoff, "Pourichekevich et l'évolution des partis en Russie avant la révolution," *Revue de Paris*, V (October, 1923), 741. 5. Pares, *Memoirs*, p. 396.

The friend commented how he was reminded of John of Gaunt who once thought of England in the same way; Russia was "now leased out like to a tenement or pelting farm."

Now was born the desperate plot to save the monarchy by destroying this infamous thing.⁶

§

Prince Felix Yusupov was a member of a rich and distinguished family descended from Tatar chiefs who had served in the Golden Horde and at one time had ruled the khanate of Kazan. He was a remarkably handsome youth of somewhat precious habits whose frivolous and dissipated life had won him much notoriety. He had attended Oxford for three years, toured the continent, and had done all the things that young lords were supposed to do—and later hinted that he had done some unusual things most of them did not do. But that was the sole measure of his fame. In 1914, after overcoming strong resistance in the imperial family, he married the daughter of the Grand Duke Alexander Mikhailovich, Princess Irina, widely regarded as the most beautiful woman in Russia. In spite of having everything, including his beautiful princess and the vast yellow Yusupov palace on the Moika Canal in Petrograd, he still was adrift.

The war had not changed his life; it remained as aimless as ever. As an only son he could technically avoid military service, "but I soon found it impossible to go on living a life of ease when all the men of my age were at the front." Nevertheless, he went on living a life of ease. He enrolled in a one-year course of officers training at the exclusive school of the Corps des Pages—and never got to the front. There was much grumbling in the country about the way in which many of the youth of the imperial family and their circle did not really enter military service. Such talk, along with the pointless life and lack of achievement, may have touched the conscience of Yusupov and made him yearn to do something worthwhile. He was swept up in the talk about the power of Rasputin.

They had met briefly in 1909 when Rasputin was still a newcomer in St. Petersburg. Madame Golovina and her daughter, who were fervent

---

5. Pares, *Memoirs*, p. 396.
6. Felix Yousupoff, *Lost Splendor* (New York: G.P. Putnam's & Sons, 1953). Purishkevich, *Dnevnik*. K.W. Hiersemann, (ed.), *Originalakten zum Mord Rasputin* (Leipzig,

admirers of the *starets* and were using their salon and teas to show him off, introduced Yusupov to him. Apparently a friendship did not grow between the two men. In 1916 Yusupov's mind turned back to Rasputin, whose name he was now hearing all the time. His mother had become a leading foe of the *starets*.[7] She had personally seen the empress and had protested his presence at court, but nothing happened. Then after a long absence at court she returned for one last try. Alix received her coldly. She believed that the Yusupov clan were among the strongest denouncers of the tsar and were involved in talk that Alix regarded as seditious. The father of the prince, as governor of Moscow, had tried to use the city as a rallying point of anti-dynastic forces. Besides, who were they to talk of depravity? The empress had expressed sharp disapproval of the conduct of their young prince. The princess poured out her warning of the danger of Rasputin's presence. Alix listened stiffly, and when the audience ended she stood up and said, "I hope never to see you again." The Yusupovs were close to Alix's sister, the Grand Duchess Elizabeth, who was living in Moscow as a nun. Elizabeth saw her sister and pleaded for the dismissal of the *starets*. After the interview she arrived at the Yusupov palace in a stunned condition, saying, "She drove me away like a dog. Poor Nicky! Poor Russia!" Everywhere Felix saw the wringing of hands and heard the laments, the cries for someone to do something about the terrible *starets*. Once again people raised the cry that had been popular in the revolution of 1905, "We can't go on living like this." The heated atmosphere now began to have its effect on the disturbed conscience of Yusupov. Here was his chance to do something, to achieve great things at one easy blow. He would be the one to remove Rasputin and save the country. He admitted he had an aversion to politics and did not understand what was happening in the country. In addition, his knowledge of the facts about Rasputin was faulty; nevertheless, he decided that he was destined to rid the country of the evil *starets*.

He arrived at this conclusion independently, but others were thinking the same thing. When he discussed the possibility of a plot with them he found them full of hatred for Rasputin but not eager to join him. However, Rodzianko, who was his cousin, encouraged him to go ahead, saying that he would join him if he were younger. Yusupov's casual way of approaching friends to discuss the proposed murder spread word about

---

1928).

7. "K istorii poslednikh dnei tsarskogo rezhima," *KA*, XIV (1926), 227-49. "Pisma D.P. Romanova k ottsu," *KA*, XXX (1928), 200-1. "Iz dnevnik A.V. Romanova za 1916-1917 gg," *KA*, XXVI (1928), 187-92. S.P. Beletsky, Grigorii Rasputin (Petrograd, 1923), p. 12.

what he was up to. His friend, the Grand Duke Dmitri Pavlovich, like Yusupov one of the gilded youth, joined him at once. The common desire to do something spectacular brought them together. Next, Yusupov turned to a wounded soldier, Captain Sukhotin, a friend of the duke. The soldier agreed to help.

Yusupov then began to stalk Rasputin. Through the Golovin women he renewed his acquaintance with the *starets*, who was glad to see him. Rasputin had few friends and many dangerous enemies among the aristocracy and those close to the throne, and therefore was eager to cultivate the prince. At this meeting Rasputin as apprehensive and restless. The air was thick with plots against his life, and he lived in constant fear of assassination. Suddenly, after the inevitable phone call, he left the apartment. Yusupov knew he had to be careful and he wanted the *starets* to make the next move. At this time he probably baited the trap by offering to eventually introduce Rasputin to Irina, who, however, was safely hidden in Crimea. Rasputin could not resist the promise of a meeting with a pretty woman. When his call came Yusupov was asked to bring his guitar so that he might play and sing gypsy songs for the *starets*. At Golovina's the prince was treated to the usual evening with the *starets*. It began with an embrace and Rasputin calling him "My dear boy." Yusupov felt sharp revulsion but said nothing. Rasputin then bragged how he had yelled at the tsar and tsaritsa, demanding things of them and threatening to leave if they did not listen to him. He also delivered a senseless religious harangue. The two women listened ecstatically although the visitor could make no sense out of the raving monologue.

By repeated visits he gradually became intimate with Rasputin and disarmed him of his suspicions. But when the time arrived for the killing, Yusupov hesitated because he didn't know how to go about it, and he became aware of the risks a lone murderer accepted. It was necessary, he concluded, to call in others for help. The plot at this stage had a representative of the officers in the army, Sukhotin; a member of the imperial family, Dmitri; a member of the high aristocracy, Yusupov; and one politician, Rodzianko. It needed more politicians. The widespread hatred of Rasputin and the thunder of denunciations against him had now reached a peak. Members of the dynasty openly talked of a coup. When in November there were denunciations of the *starets* in the Duma, the Grand Duke Nicholas Mikhailovich, among the spectators, was seen enthusiastically applauding. At the same time the members of the Allied diplomatic corps did not hide their feelings. One member of the Duma, Vladimir Purishkevich had, in fact, decided that Rasputin had to be killed.

Purishkevich is now a forgotten man of the constitutional era of Russian history, but in his day he was a famous political figure. During the revolution he became vice-president of the Union of the Russian People, a Black Hundred organization, but in 1908 organized his own group, the Union of the Archangel Michael. A pure and violent reactionary, he dreamed of undoing the October Manifesto and all its consequences. Although he never held any appointive office, he was active in politics, mostly as a Duma deputy.

With alarm and then with despair he had watched the decline of the monarchy. He hoped that the emperor would see what was happening and that his advisers would warn him in time so that he could avoid disaster. But nothing happened. Losing patience, he arose in the Duma and delivered a passionate two hour speech, a philippic in which he warned the country and the tsar about the need for action and the lateness of the moment. At the end there was a tumult of cheers and a large crowd surged around him to offer congratulations. On the following day Yusupov talked to him and invited him to join the plot. He agreed at once. Next, the Grand Duke Cyril and his wife Victoria met him and asked many curious questions. Purishkevich was sure they were hinting that they would soon ask him to join a palace coup.

Once he joined Yusupov preparations went ahead quickly. Purishkevich had an intuition about planning and an ability to get the job done. He was the director of a successful military medical train and had demonstrated his abilities as a manager. Yusupov had by this time already approached the Octobrist Maklakov for help but had been put off. Nevertheless, he went ahead with plans. Purishkevich's sense of politics told him that the murder should not be done in public but should be a private affair. The political repercussions of a public assassination might upset morale among the people and this could have bad effects on the military situation. Consequently, they decided to murder Rasputin in the Yusupov palace. The prince prepared an apartment in the basement where the deed would be done. When Maklakov was approached a second time he was more sympathetic. Although he refused to become a direct participant, he gave the plot his blessing and added that he intended to be in Moscow at the time of the killing—a prudent step since the murder might trigger violent reactions from the tsar. Moscow, a center of dislike of the royal pair, might be a safe refuge. For the plotters, his allegiance was useful since he was in contact with the leadership of the Progressive Bloc. Finally, a Doctor Lazovert, who worked for Purishkevich, was brought into the plot.

The feast was prepared, the hosts were standing ready. But how could the guest-victim be brought to the sacrifice? It was one thing to set the trap, but another thing to lure in Rasputin. He was full of suspicions, and his movements had been completely unpredictable. The prince went about the normal activities of his life; however, the city was full of rumors about what he was up to. His free-and-easy way of boasting to people about the plot had spread word of it everywhere. Even some of the police were probably on to him, but Rasputin was so universally hated that they did not give him a clear word of warning. He too heard rumors, but they came piled on other rumors and warning signals and therefore did not indicate to him exactly where the worst danger lay.

Yusupov decided to sweeten the bait in the trap, so he invited Rasputin to the palace for an evening of pleasure and relaxation, promising to finally introduce him to Irina. Such a promise had to be made if Rasputin were to be lured into a strange, new place away from his house and guards. The police had warned him not to leave home. To his daughters he confided his fears that his enemies were closing in, but he was not suspicious of Yusupov so he agreed to come to the palace on the night of 16 December.

Before he arrived, the conspirators met in the basement and watched Dr. Lazovert prepare the cyanide poison that was to kill the *starets* quickly. When Rasputin finally entered, Dmitri, Purishkevich, and Sukhotin remained upstairs playing loud music on a record player while Yusupov went into the basement to provide hospitality. Rasputin was well groomed for the evening, wearing his best clothes, his hair carefully combed, and smelling of cheap soap. The table was spread with a festive display of cakes and biscuits, and the sideboard gleamed with rows of crystal glasses and decanters of fine wines. Rasputin was nervous and distracted, and Yusupov thought he detected suspicion in his eyes. After first refusing to eat or drink he quaffed some wine and gulped a few cakes—all allegedly poisoned—but he did not drop dead. The panicky conspirators lost their heads. The prince excused himself and hastily retreated upstairs to consult with his nervous friends about what to do next. None could imagine how the cyanide failed. There were whispers, questions, and a hurried exchange of views. Yusupov borrowed Dmitri's military revolver and returned downstairs. He managed to turn Rasputin's interest toward a glass crucifix, and then, slipping around behind him and with his hand trembling, he shot the victim in the back. Rasputin roared and fell to the floor; he twitched for a moment and then was still, probably feigning death. The sound of the explosion brought the others down the stairs. In the confusion someone brushed against a light switch and plunged the basement into darkness. People bumped together, there were cries and exclamations. They were all on the verge of panic. When the lights came back on again, Lazovert is supposed to have pronounced Rasputin dead of a wound near the heart. Then they withdrew to the main floor. Later, as Yusupov told the story, he was alone with the body and could not detect a pulse. But suddenly the left eye opened, then the right eye opened and Rasputin stared at him, the grimace on his face showing malevolent hatred. In an instant he sprang up and tried to strangle the prince. When Yusupov broke away Rasputin staggered upstairs and fled the palace. Then the alert Purishkevich grabbed the gun and chased him into the courtyard. There was a fusillade of four shots and Rasputin crumpled into the snow. It was 2 A.M. Later in the night the body was taken and thrown into the Neva River.

§

In the next few days the capital was excited by the news of Rasputin's end. Most people had a good idea of who the killers were and even knew details of the plot. The newspapers, operating under the restrictions of wartime censorship, carried reports that "Since yesterday a notorious person of high importance has been missing from his usual haunts."

In the few months of life remaining to the regime there was no official investigation of the killing. Several police officials quietly gathered facts and asked questions of witnesses and of participants in the plot. Almost at once the police learned the whole story. However, the presence of the Grand Duke Dmitri among the murderers put the entire matter under the direct jurisdiction of the tsar, since only he could proceed legally against members of the imperial family, who were technically outside the control of imperial law. Nicholas was fatalistic. What was done was done, and no investigation could undo it. After Rasputin's death he had little interest in an inquiry and could not imagine what good it might accomplish. It might only exacerbate feelings in a bad situation for in the Rasputin affair there were elements that could make an investigation dangerous to the regime. Nicholas knew well what his relatives were thinking and talking about. They were contemplating revolt; their palaces seethed with plots, and talk was daily growing more desperate. Most expected the removal of Rasputin to improve things, and therefore they grew quiet after the killing. If Nicholas listened to his wife and moved with vigor against the young duke and the prince he might unite the ducal families with the aristocracy and thus cause the plotting to explode into activity.

Even an official investigation might not have gathered all of the facts, for only one witness—the principal killer—was present in the basement during nearly all of Rasputin's last moments.

For this reason Rasputin's death, like his life, has been surrounded by mysteries. Some of the discrepancies in the story of the murder are still puzzling. Legend has it that when the body was recovered from the river, the hand was in a raised position and the fingers as if Rasputin, under the ice, had made the gesture of a blessing. Naturally, the lungs contained water, indicating that the body had been submerged while still alive. Add the claims of Yusupov to these beliefs. Rasputin was supposed to have consumed large amounts of quick-acting and deadly poison, a bullet had passed through his body, his heart stopped beating, and yet he was able to get to his feet and almost strangle his assailant. Then he was supposed to be alive when four more bullets had been fired into him and when he was plunged under the icy Neva.

Police examinations of Rasputin's blood did not support all Yusupov's claims. It revealed some traces of alcohol; this agreed with Yusupov's statements that Rasputin drank two varieties of wine. But there was no poison.[8] As to the legend about the posture of the arm and the water-filled lungs, the police said that Rasputin died from loss of blood before his body was put in the river.

It must be remembered that in recounting the story of the fatal night, Yusupov was not only boasting of his deed, but also exonerating himself from blame. He was writing an apology for murder, trying to justify in the strongest terms the thing he had done. The tsar's courts, laws, and justice were swept away with the refuse of history in 1917, but the court of popular opinion still remained, and before it the prince wanted to present his best case. Therefore, Rasputin was portrayed as a demon with superhuman powers of survival. No common mortal could run the gamut of poison, bullets, and water. Indeed the last moments were supposed to reveal in final horror the diabolic

---

8. Hiersemann, pp. 8-9.

powers of the creature. In describing the last struggle when Rasputin, his shirt stained with blood, leaped at the throat of his foe, Yusupov actually wrote that he fought with Satan. Who could be blamed for killing such a monster? But Rasputin was the Devil only to the extent that he could survive one shot; five did him in. The truth is we can never really know all the facts about what happened in the basement on the night of 30 December 1916. But Rasputin did not consume poison nor was he alive when he was stuffed under the ice.

§

For Rasputin the long journey from distant Siberia and obscurity had ended. He was buried with a priest and a few attendants offering a service at his graveside. But in a way he took one victim into the grave with him. He had often hinted, when safely out of hearing distance of the palace, that the fate of the dynasty rested in his hands, that if he withdrew or was destroyed, the dynasty would be doomed. His fate and its fate were linked together so that when he was buried it might be said, in the words of John Donne,

> Both rob'd of aire, we both lye in one ground,
> Both whom one fire had burned, one water drown'd

APPENDICES:

Author's Postscript

The land around the village not only nourished the ancestors of Rasputin but also gave them their name. All the townsmen were referred to as the people who lived at the crossroad (*raspute* means crossroads) where the Tiumen-Tobolsk highway joined the turnpike coming down from Verkhoture. Later, when they came to adopt family names, in keeping with the general custom of peasants in many parts of Europe, some villagers naturally called themselves Rasputins—that is, the people who lived at the fork of the road. Peasants and serfs often took names referring to prominent features of the land in their neighborhood. (For instance, in England names like Hill, Ford, Lake were adopted in this way. Genealogists refer to these as local names). Gregory Rasputin's daughter said that her grandfather often told her that many people of Pokrovskoe were called Rasputins "when Yermak sent his ships."[1]

When Gregory became famous in St. Petersburg his enemies claimed that the name Rasputin had been coined to fit him alone and that it referred to his immoral behavior since *rasputnik* means a rake or libertine. A second school of thought said his name was connected with the word referring to the dark and stormy season of the year (*rasputitsa*) when all roads were turned into quagmires. The problem of names is a complicated one, but Rasputin's enemies treated it light-heartedly.

---

1. Rasputin, *My Father*, p. 26.

Playing slanderous games with names was a political sport in the capital. Pobedonostsev himself was also the object of such sport.[2] Rasputin was aware of the harm these accusations could accomplish and toward the end of his life he sometimes used the name Novykh (referring to "new") to escape the scandalous associations of his real name.[3] The origin of this alias is uncertain. Some said he was once greeted by Tsarevich Alexis, "Here comes the new one"—that is, the new visitor.[4] At the end of 1909 he was signing telegrams and letters "Gregory Rasputin the New" giving the name a mysterious and somewhat religious air.[5]

Some additional facts make credible the claim that his name had nothing to do with his morals. For one thing, it was often encountered in western Siberia, where prominent features in the flat landscape were scarce and where people could be easily associated with a few well known features of a river or a road. Peasants also took names from towns. For instance, Rasputin's wife, surnamed Dubrovina, probably came from a neighboring village of Dubrovin. Rasputin's name certainly was not invented to fit him alone. Not only was it met widely in Tobolsk Province, but also it was found in the city directory of St. Petersburg for 1914. The other Rasputins were not related to Gregory. In addition, the tsarist police officials, A.V. Gerasimov and Alexander Spiridovich tell in their memoirs of a plot in 1908 to kill the Minister of Justice and one of the grand dukes.[6] This conspiracy was led by Anna Rasputin, no relative of Gregory, who with her accomplices was executed in a celebrated incident later dramatized by Leonid Andreev in his novel *The Seven Who Were Hanged*. ▪

---

2. [Editor's note:] Martin intended to find a better word than "sport" here. The name *Pobedo-nostsev* means literally "victory to the bearers"—bearers of *power*, one would say, since he was known for a hatred of democracy. Pobedonostsev allegedly declared a belief that the church and monarchy, as bastions of Russian culture, ought to be "frozen in time."

3. In a letter of April 9, 1916 Alexandra pleaded with her husband to send Easter greeting to "Novy, Pokrovskoe." Also see "Telegrammy Grigoria Rasputina," *Byloe*, nos. 5, 6 (1917), 228-30.

4. Spiridovich, *Raspoutine*, p. 73.

5. Trufanoff, *Monk*, p. 103.

6. Guerassimov, pp. 191-93. Alexandre Spiridovitch, (Spiridovich) *Histoire du terrorisme russe, 1896-1916* (Paris, 1930) pp. 508-9.

From "Road to Court"

Here the author takes especially to his imaginings. Reckoning based on what we can know, he speculates about what certain characters might have known, overheard, signified or experienced of the events as they lived through them. He does this through Fr. Peter, a cleric, and the church itself. The controversy of Philippe & Encausse at court is discussed at greater length, illustrating the contemporary religious context and precedent for Rasputin's role before he arrived.

No footnotes were provided for these earlier drafts of chapters, those that appear are editor commentary.

One day in 1902 the village priest of Pokrovskoe, Father Peter, sat down to compose a letter to his superior, Anthony, Bishop of Tobolsk. With slow care he wrote out the formal salutation, then paused and sighed heavily as he thought of the painful words he was about to put on paper. In this letter he would show the bishop evidence that Pokrovskoe harbored members of the hated and feared sect, the *khlysty* (Flagellants), who in their insidious and cunning way, the priest explained, tried to lure Christians into heresy by offering them a new and loathsome faith that mixed sexuality and religion. Fr. Peter reflected briefly in order to review the evidence for the last time before recording it in the official report.

Several months ago the Wanderer Gregory Rasputin had returned to the village for one of his visits. He announced at once that God had called him to help others achieve what he had achieved—a life without sin. At first the priest was suspicious of this lay apostle who for many years had shown an interest in religion. Conversations with Gregory had convinced him, however, that the man's convictions were Orthodox although somewhat crude. Nevertheless, some of the villagers grumbled about the wanderer's strange words and actions, and eventually their complaints revived Fr. Peter's concern. When he heard that the new-found holiness of Gregory had not rid him of his old vice, drunkenness, Fr. Peter decided to begin watching the self-styled Man of God more closely.

The stirrings of curiosity and a growing fear about Gregory's intentions caused the priest one evening to accept an invitation to visit the Rasputins.

He rated them an affluent family. Although they had been turbulent and had a reputation for trouble-making, they were clever and ambitious. Lately their affairs seemed to be going well. During part of his life Gregory had spent much time away from the village, but recently he had been taking only short trips away from home. He claimed that he had visited important people in Kiev and Kazan, and boasted of friendships with churchmen and businessmen. Fr. Peter had ignored the claims although he was impressed by the presents of expensive icons, painted plates, a fur hat, and the elegant boots that Gregory showed off.

When he approached the Rasputin house from a distance the priest noted that it looked like an average peasant dwelling with an iron roof. Its side was turned toward the roadway and a tall painted fence enclosed the mews and the buildings with stalls and sheds that housed horses, cattle and wagons. The first signs of wealth appeared in the yard where a number of black pigs wandered about in company with ducks and chickens. The sash windows and their shutters displayed fully scalloped mouldings, and under the eaves on the front and side of the house ran a full-length border of richly carved fretwork.

At the door he was warmly greeted by Yefim, the elder Rasputin, who ushered him inside and offered tea. Here too were signs of prosperity. White cloth curtains filtered the sunlight coming into the room. Near the windows squatted wood vats containing large-leafed plants growing as tall as a man. Colored prints stood out here and there against the painted and polished logs of the wall. They contained the favorite themes of all pictures in peasant homes: shipwrecks, hunting scenes, a panorama of the Russians attacking at the Battle of Plevna during the Turkish War of 1877-78, a portrait of the tsar and tsaritsa, and a brightly colored representation of heaven where God was seated on a throne surrounded by a pink light. The eye of the visitor was drawn to a corner filled with a cluster of saints' pictures and icons framed in filigreed brass. Beneath them hung clumps of accumulated drippings from wax candles, decorations of cheap artificial flowers, and dried sprigs of wild herbs.

In the middle of the house before the large brick tiled oven—actually a room within the main room—stood Gregory leading a noisy but friendly discussion amidst a score of neighbors who were pressing around him. The priest received Gregory's respectful greetings and then moved among the people, gossiping and exchanging pleasantries. No food was served. Darkness fell and hours passed until the lamps were turned out, and Fr. Peter withdrew unobtrusively into a corner while the other guests arranged themselves in a semicircle around a table containing wooden candle holders. Gregory lit the tallow candles and then stood back in the soft yellow light. The group sat quietly for a while, each person staring at the small flame, lost in his private thoughts. Presently Gregory arose, and facing the light he began to talk softly, almost in a whisper, yet his voice carried effortlessly into every room. For a while he commented briefly on a number of pious themes, but soon he began to concentrate on the subject of pain and suffering in the lives of men. To his neighbors he offered consolation and encouragement about the hardships they faced in the world. For a moment he seemed to promise that he had the power to lighten their burdens; at that point, however, he suddenly stopped. After a pause he asked if anyone afflicted with troubles cared to reveal them to him. At first there was no response; then three peasants stood up and self-consciously each in turn lamented in detail about his misfortunes. The priest glanced at Gregory who was listening with deep interest. In the pale light his eyes burned with diamond points. When the speakers had finished he stood still, with head bowed and hands spread limply across his breast. The audience watched him in complete silence. Then as he talked in a low, soothing voice, he started to move among the group, offering words of comfort, stopping at times to place his hand gently on the shoulder of one of the peasants who had revealed his woes. Gregory's sensuous voice at times strained with emotion, and his body, usually rigid and tense, moved in flowing, graceful gestures. On the wall behind him swift-gliding shadows moved, disappeared, and in a moment were reborn out of the darkness into which they once again dissolved. Gradually, the priest became aware that Gregory was no longer talking to the group; his speech was apparently directed to some large unseen audience. As words flew from him he was gripped by an intense religious emotion. His thoughts were expressed in a strange and beautiful language combining the imagery and feeling of the Bible with the earthy language of the peasant—all blended in a fashion that overpowered the audience, left it transfixed and silent with thought.

    When he had finished, Gregory placed his hands over his eyes and staggered back in exhaustion against the wall. But even before he had uttered his last fervent words the audience began to stir, to sway slightly in unison. Then for a few minutes the priest caught the simple rhythmic melody of a song he had never heard before, and the atmosphere of the room seemed to be filled with some kind of hidden presence or power. For a while Gregory stood motionless, then suddenly he raised his arms above his head and announced the meeting was ended.

The moment the lamps were relit the air of tension and mystery in the room dissolved, the trance-like ecstasy of the guests changed quickly and once again they became the gay, noisy crowd they had been when the priest had first entered the house. In a short while good-byes were said and everyone went home.

Fr. Peter did not know how to judge what he had seen. The events of the evening left him excited but disturbed. It was impossible to deny the feelings of devotion and deep piousness of those hours. He had been overwhelmed by Gregory's eloquence and sincerity, yet there was something bizarre and incomprehensible about the events. The priest was sure he had witnessed a program that had been planned; the entire incident was like a ceremony or ritual, part of which had probably been cut short. He had an uneasy feeling that he had witnessed things that were beyond his understanding.

He could not deny that the villagers were much impressed by Gregory. Shortly after the wanderer's return people began to hail him as a *starets*. He was called a Man of God, and some referred to him as "Father Gregory." Villagers listened with close attention when he spoke to them on the streets or in the fields. Many commented on his ability to comfort the afflicted; others suggested he had a power to heal the sick or to see into the future.

Several months passed and then the priest began to hear more alarming rumors about Gregory's conduct. Instead of meeting openly with interested villagers, Gregory now held meetings in out-of-the-way places; at first in the mews behind his father's house, then in the family bathhouse and later in remote clearings in the woods beyond the village. Men who had passed these trysting groves at night reported that when they stopped to watch and listen they heard curious sounds—loud laughter of women, boisterous noises of people dancing and carousing. Then the lights and fires went out and through the trees drifted the sound of strange singing, not the hymns of Orthodoxy but the accented chants belonging to the religious music of the sectarians. Soon people were whispering that the mysterious meetings ended in drinking revels and orgies. Villagers reported that Gregory was often seen in the presence of women; soon it was known that women made up a majority of his followers. In the late morning he was sometimes met on the street obviously sick and out of temper from too much drinking the night before. There were days when he disappeared completely, and his neighbors reported that certain persons were always absent from the town at the same time, indicating that a regular group was probably meeting someplace. When questions were put to him about what went on at these gatherings, Gregory was evasive, but he insisted he was doing nothing wrong. Whenever he appeared at church he spoke respectfully to the priest. During the liturgy he was full of intense devotion. And with the rest of the people on Sunday morning after the liturgy he kissed the cross the priest held up to him. Eventually he declared that his beliefs were his own business and that they could not be understood by making them subject to public discussion.

But the priest was now looking more deeply into Gregory's activities. Inquiries among the villagers turned up some wild stories and many unprovable rumors. A few persons spoke out loudly and denounced him as a fraud, referring to his sordid past and the bad reputation that went with it. His self-styled religious mission, they said, was merely a cover for his old vices: lechery and drunkenness. They added threats, saying that if Gregory did not stay away from the women of the town he might be beaten or killed. But there were others who shrugged and said that he was a harmless and amusing quack. Some of the peasants, however, looked on him with dark dread and refused to tell what they knew about him for fear he might cast spells on them. They added warnings that if the deceitful Gregory were not expelled at once he would lead the priest's flock astray. Fr. Peter affirmed this fear when he recalled rumors he had heard about entire villages suddenly swept up in the faith of the Flagellants.

The priest began to systematically review the evidence as it came to him. Some of the facts were trifling but they might still be important. For instance, it was reported that Rasputin and his friends addressed one another as "brother" and "sister." This was reported to be a common form of greeting among some kinds of sectarians. The use of the bathhouse for meetings was suspicious too. In the average village this was the scene of much trouble. The gatherings in the woods formed a more important kind of evidence. The swampy forest, the *taiga*, was a gloomy woods full of sinister, dark places, the abode of evil spirits such as the wood sprite who hid in the densest thickets. In the *taiga* the weak or the imprudent lost their way amid the endless miles of silent evergreens and blackwaters and there they perished. Every village had its stories of people who wandered off into the pathless bogs and were never seen again. Only hunters and lumbermen went into the forest and found their way out. Peasants who had to go into it for any reason got out as soon as their work was done. It was unheard of that women went into the forest. The village established a burial ground for its dead in the forest but otherwise left it alone, preferring to leave it to bandits and wolves who dwelled there.

As he poised over the letter to the bishop Fr. Peter's mind filled with apprehension and doubt. He was oppressed by the realization of how poorly prepared he was to judge whether Gregory was or was not a Flagellant. He was aware that members of the sect called themselves Men of God; they were organized into groups called "ships" under leaders, male or female, called "Christs" who claimed to be descendants of the original prophet-leader who had first appeared about two hundred years before. Since they had for a long time lived under the threat of brutal governmental persecution they were secretive and used all kinds of tricks to protect themselves. The rank-and-file were not expected to confess their faith publicly; they were free to protect themselves by posing as respectable members of Orthodoxy and by attending church services. They did not revere the Bible, but possessed their own sacred unwritten literature which was preserved in their hymns and handed down through generations. The Men of God, as they styled themselves, believed salvation was assured them, and therefore they could not sin. All Russians, educated and uneducated, were convinced that the Flagellants were promiscuous and debauched, and that their religious services usually ended in a wild melee of bodies writhing about in a mass orgy.

The priest knew of no simple test which would determine if Gregory was a Flagellant. The information he had gathered about Gregory could not conclusively brand him a member of the sect; some of the information was actually contradictory. For instance, the Flagellants were not drunken; moreover most sectarians were addicted to cleanliness, something that obviously did not interest Gregory. Fr. Peter was sure only of what he had seen in the Rasputin house: a bowdlerized version, he was convinced, of some kind of blasphemous ceremony that Gregory later on probably conducted secretly in a different form. The priest suspected that he could accuse Gregory of nothing more than general sectarianism.

It would be necessary to make this charge to the bishop. Then the formal investigation would be set in motion by the authorities who would soon appear in the village in order to gather facts. Fr. Peter hoped that their appearance would force Gregory to leave. In that way the entire matter would be settled. But the priest knew that if he made the general charge of sectarianism then months or perhaps a year might pass before the officials acted. They had learned from experience that such accusations were often made by panicky priests who were forced to deal with harmless peasant cranks. However, in the present case if the government responded with its usual slowness the situation in Pokrovskoe might get out of hand. But if Flagellantism were suspected, the authorities could be counted on to act almost at once since this cult was considered especially dangerous. It was reputed to carry out vigorous missionary activity, and if it were not stamped out immediately it might attract the support of the peasantry in the village. In addition, the political officials believed that the Flagellants tried to subvert civil society and were as dangerous to the state as to the church.

All kinds of thoughts passed through the mind of the priest. The appearance of the sect was a menace to his interests. An important part of his income came from pious church folk who paid him for many services such as blessing fields, baptizing, saying special prayers over cattle, etc. Moreover, if he were remiss in issuing a warning he would get into trouble with the consistory, the powerful council of laymen who controlled most diocesan affairs. The bishop too would be angry with him, for the bishop always blamed the local priest when sectarianism flared up. When the authorities were aroused in such a way it did not take long for the district priest-inspector, a much-feared snooper who made life hard for pastors, to come and begin his questions and threats. Tobolsk would forget Fr. Peter's carelessness only after payment of a large bribe to the consistory. If this were not paid he might be put under inhibition for six weeks, and during that time he would not be able to practice his duties and would receive no pay. His family would go hungry or be forced to beg among the peasants.

This was a moment of tragic choice that a village priest sometimes had to face. He confronted a self-styled prophet who might still be induced to work inside the church. Get rid of the so-called Man of God, but risk popular clamor. Or take him, work with him, and even make good use of his popularity with the people, and then try to subtly bend and change his heretical ideas. Fr. Peter decided to avoid the risk of permitting a Flagellant to live in the village; he would remove responsibility from his shoulders by charging Gregory with membership in the sect. He felt justified in these suspicions; if he were wrong about the kind of sectarianism involved in the accusation no harm would really be done. Then it would be the task of experienced bureaucrats to decide about the fine points of words and deeds that distinguished one kind of cultist from another.

How would the officials evaluate the information? Fr. Peter was a poor village priest; economically he was not much better off than the peasants he served. He had some education, it was true, but he was in many ways not much more sophisticated than his people. The officials had more experience and training, and to them the Pokrovsoe incident and the frightened pleas of the cleric were covered with danger signs.

First, the report went to the office of the provincial governor and to Archbishop Anthony of Tobolsk, who passed it on to the office of a typical church bureaucrat. We can imagine the clerk who received it: a bored layman steeped in contempt for all the problems that troubled the hearts of clerics. The report on Rasputin may have given him a moment of diversion, for he could picture the sectarian's antics. Perhaps these were the revels of Midsummer Night or St. John's Eve (*Rusaliya*, the Russians called it), the time of hot weather madness when peasant men and women temporarily lost their fear of the deep woods and gathered in a remote clearing to build a bonfire and to indulge in shameful dancing and chanting. Later, when their Dyonisiac rites struck them with intoxication they began to shout and scream as they leaped high over bushes and fallen trees, and as the revels grew wilder they performed unspeakable acts in honor of the Corn Maiden. At some of these Saturnalia Satan was supposed to appear in the form of a rutting black goat who mounted a decaying tree stump while his disciples, dressed as serpents, toads, and wolves, danced wildly around worshipping him in ecstasy. At dawn when their frenzies had exhausted them and all wantonness was drained from them, their bodies grw limp and they collapsed in sleep. Such things had been depicted in a popular book written a generation ago by Matvei Khotinsky, *Witchcraft and Mysterious Phenomena of Modern Times*. It was crammed with fascinating legends—some based on facts—describing the forest nights. A famous composer, Mussorgsky, had even composed a piece of program music portraying the night when witches rode to join a Sabbat on a hill near Kiev.[1]

For a while the official savored these facts with some friends who snickered with him. When they considered the matter seriously all the possibilities of the Pokrovskoe case seemed bad. Peasants were known to revert sometimes to heathen ways, to return to the past when pagan Slavs had no temples but gathered in sacred groves to set up idols and worship them. Experience taught that secret ceremonies plus secret chambers equalled trouble. Add the possibility of sex and drink and there was more trouble. These might not be things to jest about. Perhaps a new cult was being born. One could not predict where such groups might appear; therefore, one could not be too careful. Some of the evidence in the Rasputin case was disturbing. The bureaucrat knew that K.P. Pobedonostsev, the Over Procurator of the Holy Synod, had been holding rallies of churchmen lately, warning them that the sectarians were now unusually active. Pobedonostsev could make or break careers, and he surely would be harsh toward anyone who had not taken his warnings seriously. His sharp eye was known to follow the matter of sectarianism down to the village level, and he was known to make personal inspections of "infected regions." Recently he had complained about the Theodosians whose prophet I.A. Kovylin, had said "In paradise are many sinners, but no heretics." The Theodosians were also rumored to be horse thieves. All this sounded like Rasputin.

The clerk recalled a well-known case of nearly one hundred years ago when one of the closest friends of the tsar, P.A. Golitsyn, became a self-styled religious leader among dignitaries of the court. He was reported to have held secret prayer meetings that were followed by wild parties. Only the exertions of the most powerful churchmen were able to exorcise his evil influence on the tsar himself. The daring of such foes of the church was amazing. Just a short time ago he heard of a case in the Crimea that had disturbed official circles and caused the Holy Synod to issue a flurry of warnings to authorities calling for increased alertness in rooting out heresy and sectarianism. In this incident a group of impudent Stundists (Baptists), who were regarded as a mild and moderate sect, sought to proselytize the faithful by claiming they were supported by the empress and the Grand Duke Alexander Mikhailovich.

The bureaucrat decided that discretion in this case was less important than caution. Besides, he did not know what to make of some of the evidence. He knew little about Siberian peasants, he dreamed of an assignment in St. Petersburg someday, and therefore he had little desire to study peasant ways. The suggestion of misconduct in the bathhouse he had to take at face value, for instance. He knew almost nothing about this peasant institution. He did not know that *khlysty* did not drink; this fact alone would absolve Rasputin from membership in the sect. In addition, the *khlysty* were not guilty of orgies. This charge was always made against them but it was not true. It was a typical slander of a religious minority, and it was widely believed. The bureaucrat knew that laws against the *khlysty* dated back to the time of the Empress Anna. A quick consultation

---

1. If one prefers a citation here, UBC student Kelly Lyn Dore, in her (December 2008) thesis on the subject, cites Edward Reilly, "The First Extant Version of Night on Bare Mountain," in *Musorgsky; In Memoriam 1881-1981*, ed. Malcolm H. Brown (Ann Arbor: UMI Research Press), p. 138.

of the *Complete Collection of Laws of the Russian Empire* revealed the official statute: ukase of August 7, 1734, number 6613. He had read many attacks on them in the ecclesiastical journals. He recalled an article in the *Russian Messenger* of thirty years ago which had mentioned this charge. He did not know that even churchmen were beginning to admit the accusation was not true. If he had bothered to read the *Missionary Review* of January, 1901 he would have seen that the church officials were willing to concede that such charges against the Flagellants were wrong. But he was not interested much either in the church or in peasants. He had enough of their bath-houses, fanatical sects, and animal-like lust. He wished he were in St. Petersburg or Moscow. He recommended an investigation.

§

The brains and leadership of Orthodoxy was concentrated in St. Petersburg where all the complicated apparatus for controlling and directing the day-to-day operation of the church existed. In hundreds of offices leaders argued, consulted, reached decisions, and issued orders to more than 100,000,000 faithful and 200,000 clergy, spread out over millions of square miles of the empire.

...Tempests might buffet it, but could not destroy it. In 1905 the government and church were to be caught unawares by the shock of revolution, but the disasters that struck the church in the twentieth century did not come as a surprise to a number of the clerics. They realized what an observer might see when he looked at the church. From afar it appeared solid, majestic, surrounded with an air of tranquility and venerable age. The only sounds coming from it were those of pleasant harmony. However, when he came closer he saw that the structure sometimes shook with tremors; closer still and one could begin to see cracks in the walls. He could hear discordant sounds, laments and complaints, voices raised in angry tones, and the clash of demanding cries. Smoke and dust rose from unseen struggles raging inside the walls of the unhappy building.

...Before the church could rise, before it could hope to solve its own problems, it had to cast off the control of the government. But in the meantime the problems of the church kept growing.

The most striking problems resulted from its association with the government. In the past tsarism had strangled patriarchs and killed bishops, and it treated the lesser clergy with contempt, reducing them to the status of a closed, self-perpetuating caste. The man who personified strong control was Pobedonostsev, perhaps the most hated man in Russia, who presided over the Synod from 1881-1905. He had been a personal friend and tutor of Alexander III and a teacher of Nicholas II. It was known that he was consulted on state affairs. As a result he was regarded as one of the principal architects of the policy of reaction that the government had followed since the start of his reign. It was also widely believed that he controlled Nicholas II. A popular picture showed him with arm outstretched imperiously ordering Nicholas to sign a famous and hated ukase of 1894 in which the new young tsar reaffirmed the policy of repression that had guided his late father. Pobedonostsev's power came from his personal relations with the sovereigns. From his high position as Over Procurator he brooded and watched like an alert hawk over the government and the church. However, he was not capable of feeling any genuine religious emotion or sentiment. To him the church was only a means of social control, a convenient weapon to crush the seething spirit of rebellion in the country. But he did not have much confidence in the church even for this purpose. He was a clever, learned man who had been a respected professor of jurisprudence and a protector of Dostoevsky, whose views resembled some of his own.[2] He hated all assumptions of the liberal, optimistic spirit of progress that dominated the intellectual world of his time. Pedantic and narrow, angry and shrewd, he had an aloofness from his era that permitted him to see things in his own dour, carping way. But he was not a simple-minded reactionary; although he dreaded the future he was not impressed with the past. He was a debunker who sometimes raised valid questions about cherished institutions and who insisted on asking if men were not fools for putting all their trust in courts, laws, a free press, and parliaments. But nobody listened to him. So he plodded his own way with power to comfort his loneliness. Out of pages of history books his full face portrait still stares at the reader; with defiant curiosity he peers over small steel rimmed eye glasses that rest low on his nose. The large domed head with the long upper lip and small chin make him look like a petulant monkey.

But this was the man who ruled the church, unconcerned about public opinion. No appeal by enlightened clerics reached him because he surrounded himself with his own men, lickspittles all.

Although Pobedonostsev was the symbol of the church's woes he was not the cause. There had been even worse Over Procurators—fools and atheists, for instance. Count A.N. Protasov (1836-1855) was a hussar officer, an elegant man-about-town who swaggered into the synod meetings dressed in his red uniform, dangling his off-the-shoulder jacket in debonair fashion, with sword and spurs clanging as he walked through the room full of bishops. When he spoke his words brought no joy since he expressed admiration for certain ideas of the Jesuits and also proclaimed that one of the main jobs of the clergy was to arouse patriotic fervor and xenophobia among the masses.

The church groaned under other burdens too. Some clerics worried over the loss of the intellectuals, many of whom were now indifferent or hostile to Christianity. Other clerics felt that the church did not need them since it had the solid support of the peasantry. These ecclesiastics were concerned with the first attempts to reestablish contact with intellectual circles late in the nineteenth century, attempts which ended in the creation of the Religious Philosophical Society in 1901 and went on to the organization of other study groups. The laymen in these seemed to have already passed beyond Christianity to a religion that was very learned and syncretic, one which cultivated recherché literary fads and a bizarre naturalism that was close to pantheism. Such organizations did not look like the gateways through which the lost sheep would be lured back into the fold.

Those not concerned with intellectuals were apprehensive about another problem, the growth of non-Orthodox Christianity. In the 1870's Protestant mission groups under Lord Radstock and Colonel Pashkov had distributed bibles widely, and had held meetings and hymn singings in the streets, propaganda devices unheard of in Russia. Despite strong government counter-measures, the Protestant threat, it was feared, continued strong. Mission societies in western Europe were known to be intensively training workers to go to Russia. In addition to this threat, sectarians were growing more bold and daring with every day. They were showing less concern with hiding their views and more willingness to prosletyze aggressively and even belligerently. They seemed to think that all Orthodox were convertible and, it was rumored, they were preparing plans for a large missionary drive in the near future. They were no longer a catacomb church.

Early in 1901 the novelist Leo Tolstoi, who had in his later years fancied himself a great moralist, was excommunicated. Though it was an unusual step for the church to take, there seemed little choice after years of provocation on the part of Tolstoi. The public reaction was surprising. The church and government were thrown completely on the defensive.

---

2. Pobedonostsev befriended the author Fyodor Dostoevsky as a patron and advisor; after Dostoevsky's death, Pobedonostsev curiously took credit for shaping the author's understanding of church matters (discussed differently above, on p. 36).

At home and abroad no one seemed interested even in listening to the case of Orthodoxy. This episode dramatized the situation of the church; it was plainly under attack from many enemies. The problem was how to defend it.

[...]

The early history of the Romanov dynasty was closely tied to the church, to which the dynasty partly owed its crown. The father of the first Romanov tsar was Philaret, the Patriarch of Moscow. Loyal clerics who wanted to avoid joining the political opposition—which in Russia meant revolution—looked with apocalyptic hope to 1913 when they thought the tsar on the tercentenary of the dynasty would call a *sabor* to match the one that had given the throne to his ancestor in 1613 in the cathedral of Kostroma.

The people who wanted a *sobor* realized that such a gathering would seek to provide for the restoration of the patriarchate that had been ended by Peter I early in the eighteenth century. The hierarchy were quietly but passionately discussing the idea but the reformers could not agree on the method of selection. Some wanted democratic participation of all the clergy, others said only the leaders of the hierarchy, in the Roman fashion, should pick the new patriarch. So the church was split on how best to set its won house in order. It could not begin a program of action because its ideas and organizations were not ready. But when the time came to act the church might have to contact the masses and enlist their support. This was a time when the political opposition to the government, sensing revolutionary opportunities in the future, was forming political parties. The church, which also had a case to make against the established order of which it was a part, was preparing to present its demands in an organized way. Like the revolutionaries, the church faced problems in contacting the masses, something it had not been permitted to do for centuries. In such a situation peasant holy men who could speak the language of the people might be useful.

Although the moment required the utmost caution in words and deeds, from time to time impatience seized some ecclesiastics and they rushed ahead. Once when Pobedonostsev was ill and could not attend a meeting of the Holy Synod, Metropolitan Anthony, carried away by the unexpected opportunity, raised the issue of the need for freedom. When the revolution broke out such incidents became common, and the desperate Pobedonostsev tried to call in the support of the police official D.F. Trepov, who had been charged by the tsar to crush the revolution. But even this denizen of reaction knew that cossacks and whips could not deal with the church; besides, Pobedonostsev was on his way out, the tsar had grown weary of him, so pleas and warnings went unheeded. Still the Church did not go free.

Some of the most important clergy were to be found in the St. Alexander Nevsky Monastery and its Ecclesiastical Academy. The monastery was situated on the banks of the Neva River that flowed through the capital. It stood at the end of the main street of the city, the Nevsky prospect, on the spot where, Prince Alexander of Novgorod was said to have saved Russia by defeating the Swedish invaders in 1240. At the start of the twentieth century it was located in a working class district. Its burial vaults held the remains of some of Russia's most illustrious men: Lomonosov, Glinka, Karamzin, Dostoevsky, Tchaikovsky, and others. Two hundred years earlier the Emperor Peter the Great, desiring to add lustre to the city he had recently founded, erected the monastery and spared nothing in making it an imposing edifice. From the town of Vladimir, near Moscow, he transferred the bones of Alexander and himself acted as steersman and pilot of the barge bringing the remains of the saint, which were placed in a casket of 3600 pounds of silver. On the walls he hung many treasures of artwork including paintings by Van Dyck and Rubens. About 1800 a school was founded in the monastery; it was soon raised to the rank of a full fledged academy, and at the same time the monastry was made a Lavra, placing it first among the four leading monastic institutions in the empire.

[...]

A few weeks after the [royal couple's lavish 1901 visit to France] a newspaper, *Echo de Paris*, which was secretly subsidized with Russian money (from the so-called "reptile" or "serpent" fund) began a series of articles by an anonymous author, "Niet," who warned Nicholas that he was in danger because his enemies were plotting a palace coup. "Niet" concluded with some general advice for the tsar: if revolution came it would be fomented by the upper classes, not by the people; trust the people, for they would never betray a tsar who had faith in them. If the peasants revolted it would be when the tsar was gone. Both the French and Russian governments at once began an investigation to discover who had written the article. This was not easy to do because Paris was a seething arena of intrigue for many branches of the Russian government, which could say and do things there that were not permitted at home. Most of these feuding elements gathered around the secret police, the Okhrana, whose headquarters for foreign activity was in Paris. The police were well provided with uncontrolled funds for buying the loyalty of newspapers, votes of politicians, and favorable speeches by statesmen. But they also had money for darker activities: slander campaigns, blackmail plots, and forgeries. The article became the center of a storm raised by an inevitable bureaucratic report. It was hard to know who ordered the report. Some said D.D. Sipiagin, some said Witte.

    The Prime Minister Sergius Witte feared the articles were aimed at him since he had expressed a desire to see the tsar's younger (and more pliable) brother take the throne. He ordered Jules Hansen, in charge of press subsidies, and P.I. Rachkovsky, head of the foreign service of the police, to investigate the articles. The police official carried out this task ably. He accused Encausse of being the author. Encausse dabbled in conservative politics, although most conservatives in France who had heard of him disliked and feared him. He had a reputation as a writer of books on magic, occultism, and spiritualism. He was also closely connected with Philippe whom he called his master. Rachkovsky later rounded out his work with a report on Philippe which he sent to P.P. Hesse, Commandant of the Palace. Hesse, a charter member of the group of investors developing the notorious Korean timber concessions, was one of a clique trying to push the Russian governmnt into an aggressive policy in the Far East, something that Witte was trying to prevent. Hesse saw a chance to hurt Witte and Rachkovsky so he passed on the report to the tsar. The report contained many truths, but it was padded with misstatements, which unfortunately for Rachkovsky the tsar was in a position to know about because of his personal relationship with Philippe. Nicholas also knew that when one minister wanted to injure another he could go to the police, who were known to deliberately misrepresent materials held in their files in order to be on the winning side in a ministerial feud. Nicholas was inclined to discount almost all personal information that came to him from any government security service. Later when he was warned about Rasputin he glanced at the reports and then dismissed them, even when they were essentially right.

    In April, 1902, V.K. von Plehve was appointed Minister of Interior to replace the assassinated D.S. Sipiagin. Plehve was a believer in a hard and repressive policy as the best bulwark of the government against the rising tide of revolution, and he was also a relentless enemy of Sergius Witte who preferred a more supple approach in dealing with revolution. The police were under the control of the Minister of Interior and it was perhaps inevitable that Rachkovsky, said to be Witte's follower, should find his career ended—as it was after Plehve's appointment. But when

Plehve was in turn assassinated in 1905 Rachkovsky enjoyed a comeback as Vice-Director of Police of the Ministry of Interior. Most people assumed he had been forced to resign as director in Paris because he had dared to offend the tsar by telling him the truth about Philippe. The fall of the powerful and feared Rachkovsky had an effect on other persons a high police official V.A. Dedulin and the Empress Dowager both stopped their attacks on Philippe when they found their influence at court slipping. General Hesse, who personally disliked Philippe, expressed his feelings of disapproval but fell silent when he learned of the power of the newcomer.

[...]

Both Philippe and Encausse gave advice which it was hoped would provide the tsar with a son. By this time the empress had become a semi-recluse because of her physical ailments, probably most of them induced by her imagination. The best medical advice had not been able to do anything for her. One doctor had written that her troubles were *furor uterinus*, a minister had said privately of her ailments, *c'est une question clinique*. But none of this helped her. The couple felt isolated and almost friendless, beset by problems they did not understand in a country drifting toward an explosion. The words of the Frenchmen were one of their few sources of encouragement. Throughout their careers they had been interested in bringing mental and spiritual consolation to patients, especially those who suffered from troubles of the mind. Encausse's ideas on the medical arts were in some ways a throwback to the days when medicine had not yet emerged from magic. All his dabbling with occult lore was put at the service of healing. He sensed that the first step in treatment called for improving the patient's mental outlook by giving him hope and the assurance that there were powerful forces assisting him. Since body and soul were united, treatment of the soul could cure bodily ailments. He thought that prayer might be a legitimate form of medical treatment of disease. It would succeed when all other methods failed. For the political problems of the tsar he had some answers. He claimed that the modern world was on the verge of a great transformation in which the sun of the age of science and material progress was about to set, to be followed by a new dawn in which the sun of spiritual forces would rise and rule the world. Modern knowledge had yielded all the benefits it was capable of and now it was "gangrenous with skepticism," begetting evils such as revolution and regicide, sure signs of its coming decline. Christianity would revitalize the West. God was ready to reveal his plans to save mankind but He needed wise and faithful servants to help him carry out his plans in this world. He would seek rulers to work for Him.[3] But where could such men be found who had kept the faith, who had not let themselves be seduced by modern ideas? In his letter of 1896 Encausse pointed to Nicholas and said he was the destined sovereign-savior, the hammer of the Lord, the new Charles Martel who would turn back the forces of modern heresies now pounding at the inner gates and threatening Russia, the last sanctuary of Christian civilization. Just as Joan of Arc had rededicated France, so Nicholas would save Russian civilization and all the world with it. Nicholas was prepared to receive such ideas warmly; they sounded like the sentimental and romantic nationalism of Dostoevsky and Pobedonostsev, Nicholas's teacher.

The main problem for most observers was: who were Philippe and Encausse and what were they trying to do? Everything about them had an air of mystery. [...] However, observers knew that in January 1901 Encausse had set out for Russia bearing a letter from Delvaud, *chef de cabinet* of the Ministry of Foreign Affairs, requesting that all French diplomatic and consular agents in the empire extend him assistance in trying to carry out the purpose of his trip, which was to establish in Russia some things called psycho-physiological institutes. [...] To this day members of the tsarist emigration say that both men were really French agents trying to keep Nicholas loyal to the alliance with France.

Churchmen were disturbed when they looked at some of the people associated with Encausse. There was Saint-Yves d'Alveydre, a friend of the Grand Duke Peter. An intellectual associated with a Theosophic publishing house, he reminded his readers that in the primitive church all members—bishop, clergy, and laity—were equal and the hierarchy were elected by the faithful. Was there a danger that this man was proposing the same arrangement for the Russian church? Another one of these Frenchmen was the writer Paul Sedir who was preaching the mission of the White race in Asia. Nicholas was known to be responsive to this message. But would he listen to other things Sedir had to say? Sedir was a student of the writings of Jacob Boehme, a sixteenth century German mystic who had influenced Quakerism and Pietism but whose ideas were hidden behind a cloud of obscure rhetoric. Finally, Sedir's ideas about grace and salvation were thoroughly Protestant. The man obviously could be dangerous if his views were presented by a friend at court.

Encausse was known to be a Theosophist, a sect organized by the expatriate Russian, Madame Blavatsky, who claimed to be a relative of Sergius Witte, whom the church mistrusted. [...] both Theosophism and spiritualism attracted the support or curiosity of a wide range of important people all over Europe toward the end of the century when it became the rage and was to some a parlor game but to others a genuine way of talking to ghosts. The fad had become so widespread and pretentious that the Holy Office of the Vatican had to solemnly condemn it in a decree issued in 1898...

The most disturbing thing about Encausse was his association with the Martinists of Lyon. Martinism, a mystic, aristocratic, and Mason-like rite was organized in Paris in the middle of the eighteenth century. Lyon was one of its most important centers; from there it spread all over Europe including Russia where it recruited Tsar Paul (1796-1801) as a member. The church came to look on the rite with deep suspicion since it reduced revelation and organized religioin to a low status. Encausse had revitalized the Martinists in France, and he had set up lodges in Russia where they stressed liberty, equality, and fraternity, the slogan of the French revolution and the death cry for Russia's old regime. Hesse-Darmstadt, the homeland of the empress, was said to be an active area for Martinism. The leaders of the rite had written letters to the prince of Hesse in 1810. Some wondered if the empress had come to Russia because she was a devotee of the cult and hoped to spread it among the people.

[...]Radicals abhorred them too. A new Russian radical journal, *Liberation*, published in Stuttgart, Germany, had attacked Philippe and said that the tsar was making a laughing stock of Russia by permitting himself to be influenced by such a charlatan[...]

[...]Even the departure of both men was surrounded by the same kind of mystery that surrounded their arrival. Philippe's son later said

---

3. This bears resemblance to millenial theologies that sprouted in the United States, especially in the nineteenth century. In recent decades, as literary and religious studies scholars have begun to serious attend to the content and influence of such ideas, it has become popular knowledge that Biblical arguments were deployed on both sides of the American civil war: on behalf of the Confederacy, it was claimed that social hierarchy was a necessary part of a God-fearing society, that the secular morality of abolition was a rejection of the idea of servitude which was necessary if we were meant to serve God. A hierarchy built on slavery not only hearkened to the Old Testament, it affirmed the separateness of the sacred and profane, the exalted status of the Son of God and the type of kingdom over which one might expect Him to preside opon His return. Politics have always been entangled in this, but the alarming utility of such logic in the rise of the Evangelical Right has only slowly dawned on most scholars, who typically think of such outlandish beliefs as marginal.

that the death of a favorite daughter brought the old man back to France where he died a short time after. But in Russia there was speculation that he had convinced the empress she was soon to beget an heir. The false pregnancy of 1902 embarrassed the court and caused the healer's stock to go down. He was known to have favored an aggressive foreign policy that led to the disastrous war in Asia. Encausse had consulted his "magic numbers" and pronounced that the war would bring a Russian victory. Philippe left in 1904 and Encausse in 1906. In her letters the empress during the first world war wrote favorably of Philippe. Encausse had a distinguished record as a military surgeon during the war, and when he died in 1916 all sections of the French Press, including monarchists praised him as a patriot. ∎

Appendices

From "What must I do?"

Nicholas II was a failure. He mismanaged his reign; he was not able to deal with the political problems that cried out for changes as extensive as the Great Reforms of 1861-1881, he cast aside offers by moderate political elements to play a limited role in government, he led Russia into a pointless war with Japan, he did not take advantage of the peaceful years of 1907-1914 when he might have consolidated his regime, he bungled the leadership of the national war effort in the first World War and brought ruin to the entire tsarist system. He had opportunities to act and he threw them away, he had excellent statesmen to serve him and he ignored them or failed to appreciate their value. However, the fate of a ruler is not the fate of his country; he may fall or die but the country can go on.

The story of Rasputin's spectacular career is told against the background of the life of Nicholas and the Russia he tried to rule.

§

...When he was in his early twenties, his father pronounced him "a child" and was in no hurry to assign him any responsibilities. He made a few attempts to prepare him for rule by having him attend meetings of some government agencies. As a result Nicholas sat with the august Council of State, the greybeards of the empire, but he found its deliberations conducted by the greatest statesmen of the nation a bore, and he looked forward to the day of deliverance when he no longer had to be present at such meetings.... One day in 1890 his formal course of education ended and the happy Nicholas celebrated by writing, "Today I have definitely and forever finished studying." But some form of education went on in the rituals of court life that he now became a part of [...]

There were serious moments when he listened to lectures on history and religion, especially from his uncle, the Grand Duke Sergei, probably the second most hated man in Russia. Sergei, a deep-hued political reactionary, preached a narrow xenophobia and was a super-Russian who denounced Byzantium and the Greeks and the rest of European civilization, and proclaimed the need for continuing the suicidal policy of Russifying the empire...

Alexander III had not been educated to rule. His older brother was the crown prince but he died, so there was an empty throne and one unused fiancé. Alexander dutifully took over both and became an effective ruler and husband. However, his policies of repression and refusal to reform prepared the way for disaster. He passed on to his son a lidded cauldron seething with confined energy that was about to explode.

...Gen. Danilovich, an early teacher, instilled in [Nicholas II] the ideas that the tsar must be aloof; he must take counsel with no man after he had made up his mind. His English tutors taught him the love of exercise in the outdoors and the habits of the gentleman-king in whom friendliness and consideration for others were important virtues. He must never offend, never say no in a way that might wound the sensibilities of another—a dangerous idea in an autocrat...

...[Nicholas's flirtation with Mathilde] was ended with a conventional ruse: [his father] sent his smitten son on a trip around the world, ostensibly for reasons of state. The trip was dull. Nicholas did all the things proper for a Victorian traveller, taking in the historical sights from Italy to Japan. The official volumes describing the journey aptly breathe the ennui the tsarevich must have felt. In Japan the trip ended when he stepped on the grounds of a temple and was almost killed by a sword blow delivered by a religious fanatic.

In the summer of 1891 the flirtation was renewed [...] The petite and vivacious Mathilde was highly regarded by the critics and soon won the coveted position as *première danseuse* at this moment when the glory of the Russian ballet was about to burst on the world. Nicholas enjoyed being close to someone whose talent and training won her fame; most of the people he knew were born to their positions. There were unburdened moments in her house on the English Prospect where he gathered in dalliance with a few friends and several of the young dukes to sing Georgian folk songs, to play the guitar, and to gambol satirically through such pieces as *Red Riding Hood* with Nicholas playing the lead by wearing a basket on his head. Then it suddenly ended [...] Important events came in rapid-fire succession in Nicholas's life: engagement, wedding, death, funeral, coronation...

He was a true Hamlet paralyzed by an excess of feeling rather than thought. When he was beseiged with doubts at crucial points in his reign, he sat and pondered his dilemma, but his thoughts did not grow clearly, they were not placed end on end in any rational arrangement because his emotions took over and directed him. when his feelings had dragged him up a blind alley and when the times called for a decision—any kind of decision—he acted; with eyes closed and lips moving with the words of a prayer he wrote his decree or signed his name. Historians have tried to analyze the mind of Nicholas and they have found it difficult to make firm statements about it, but the temptation to dismiss him as a fool has for many been overpowering [...] he was not an autocrat at heart...

Nicholas admitted his deficiencies of character, he was aware of the flaws in his personality and suspected that they made him a misfit on the throne. To his fiancé he complained that no one was listening to him during his father's last hours. He was being totally ignored, no one was paying the least bit of attention to him. To his uncle, Grand Duke Vladimir, he confided that he was utterly bewildered by the responsibilities of power and could not keep peace even among the many relatives of the imperial family which under his permissive leadership was disintegrating into groups of squabbling factions. He lamented, "The whole fault is in my stupid goodness. With the sole desire of avoiding quarrels which injure family relations I give in again and again until I finally appear to be a fool, without will power or character." This small problem was a microcosm of the larger problem Nicholas faced in ruling his empire...

Witte regarded [Nicholas] with a mixture of pity and scorn. These [family advisors and government figures critical of him] did not understand the truth contained in an epigram of La Rochefoucauld: *the weak cannot be sincere.*

The wellspring of his feeling and action was his profound sense of resignation and an awareness of his responsibility for the preservation of the autocracy. There was a silent and fated air about him; whatever his destiny might be he was determined to accept it with quiet courage. His sense of resignation was so strong that his ministers regarded it as a kind of death wish. Some were puzzled that his behavior should be so feckless at moments when political common sense cried out for determined action. Others believed that if his actions were wrong at least they were consistent; they thought he was following an ideal. The Foreign Minister, A.P. Izvolsky, stressed this interpretation and told the story of a dramatic moment when he called at Peterhof in August 1906. The country seethed with revolution, the first Duma had been dismissed, Witte had stepped down as Prime Minister, all over European Russia the peasantry were in a state of rebellion. Izvolsky recounted that during the interview he and the tsar could hear the booming of guns in the Kronstadt fortress across the bay where a revolt was going on. The windows of the imperial offices rattled from the explosions. He found his master calm and relaxed, and when he referred to the gravity of the crisis Nicholas replied giving his philosophy, "Why should I worry, the will of God be done, I accept with perfect submission." The keeper of the official war journal of the government from 1914-1917, a man who was a friend of Nicholas, said that he was the worst fatalist he had ever seen.

He believed that his fatalism was really the virtue of a God-inspired endurance. The most important thing was to hold on...

Cemented by the bond of faith, his will and his power would give the strength to safeguard the autocracy which God had given him in trust. To rule like a tsar amidst revolution and war was an act of faith. He knew he lacked the majestic presence, but he hoped that if he had sufficient faith in himself as an autocrat he would show that he had gifts greater than the regal ones he lacked. The first World War seemed the great trial he was waiting for. In its crucible he sought to show what he and his country could endure. In other warring countries leaders cracked when they discovered what their acts had wrought: Kaiser Wilhelm wanted his troops withdrawn from the assault on Liége and marched eastward; the director of armies, Von Moltke, suddenly realized the Kaiser was a fool and a weakling, and as a result his own confidence deserted him; in France the stolid Pétain was unable to go over from defensive to the offensive in 1917, even though the fate of France depended on this change. The crushing responsibilities of the war and the cruel demands it put on leaders caused everywhere a numbness of executive will and ruined the careers of scores of statesmen. Men like Clemenceau, Lloyd George, and Ludendorff drew strength from the tough (even the cruel) parts of their personalities. Nicholas turned to other sources. As Russia went through defeat and suffering he knew it could do what neither Germany nor the Entente could do—it could suffer and carry on at a time when reason said there was no longer any hope. Let the allies have their matériel; Russia had other resources. Nicholas was firmly convinced that if human skill and endurance could not save Russia then God would. He dreamed of a repitition of the "Divine Miracle of 1613" when the country was torn by civil war and foreign occupation yet it suddenly revived and reestablished unity and power. Nicholas also believed that in 1812 the Lord treated the

French invaders as he once had dealt with the Assyrians. He had sent the coldest winter in a century to smite the French. He demonstrated what he would do to the enemies of a faithful tsar. Perhaps there would be another miracle, but such things came only to those who believed.

...Part of the success of P.A. Stolypin, Prime Minister from 1906 to 1911, in dealing with the tsar can be attributed to his refusal to permit Nicholas to deceive himself with excuses. But there were some excuses that even Stolypin could not turn back. Nicholas once balked at signing a law granting toleration to the Jews. He said that although the prime minister had made a strong case for the law an inner voice told him that there was danger in it. This device was always one of Nicholas' trump cards; it was he voice of his father. Stolypin could not argue with ghosts, so he said no more.

...[His unscathed encounter with crowds at the canonization of St. Serafim] made it difficult for anyone to warn the royal pair of the oncoming danger of revolution... The police constantly admonished the tsar about plots against his life. He did not believe that they originated in popular discontent. They always came from "rotten intellectuals" and fallen-away gentry. He knew that many of the salons of the government despised him and that [high] society disliked him and his wife. He was also aware that few of the members of the imperial family had respect for him. But none of these facts had an effect on his confidence in the Russian people. He believed that he was pointing to his real enemies when he said to a French ambassador: "The miasma of Petrograd, you can feel them even here at Tsarskoe Selo twenty *versts* away. And it is not from the people's neighborhoods that these odors come; it is from the salons. What a shame! What wretchedness!" [In the term "miasma"] he lumped together terrorists, liberal professional men, society ladies, and sporting grand dukes. "Opinions" here became "senseless dreams" of the feeble intelligentsia... Nicholas knew that the ambassador, Maurice Paléologue, had contact with all the wits of the capital and he would bring these words back to them...

In 1913 he had a chance to show how he felt about the upper classes. In that year there were nation-wide celebrations honoring the tercentenary of the dynasty. There were many open-air liturgies, processions of massed priests and bishops, and parades at which famous icons were displayed. Most of the events took place along the Volga, in the city of Kostroma, and in the churches in the Kremlin—all places where society did not care to go. As a result there were bitter words heard in the capital; never had the upper classes been so snubbed. They were bored with all the stories of the imperial family attending liturgies with hordes of muzhiks in unknown villages and remote towns. Nicholas returned from the festivities, spent only a short time at Tsarskoe Selo to take care of state affairs, and then went off to the usual summer cruise along the Finnish coast. The family came back in the autumn and then immediately left for Livadia, their residence in the Crimea, where they stayed until the end of the year...

He was smug in his conviction that peasant Russia was loyal to him and believed this was all he needed. The rebellious state of the countryside in the summer of 1906 had shaken him more than the news of revolts in the cities. After the revolution he adjusted his views...that although they could be stirred up by revolutionaries [peasants] were essentially loyal to the throne.

...Nicholas had to rely upon the groups which traditionally provided servants, to fill the ranks of administration... A life of faithful service... no longer won assurance of honored retirement... Old Admiral Rozhestvensky, for instance, warned Nicholas that the proposed naval expedition to the Far East in 1905 was doomed. In spite of his own opinions he led the squadron on its remarkable adventure around the world. Wounded and imprisoned by the Japanes, he was greeted on his return home not as a hero but a scapegoat. He was court-martialed and died shortly thereafter. Other servants of the tsar felt the sting of his chastisements. For years, A.A. Lopukhin, a man descended form the first Romanovs, had filled a difficult post in the government. He recruited the police officials in the capital and found men to fill the dangerous assignments in the ranks of the Okhrana, the state police. It was he who pushed the career of Alexander Gerasimov, an official who played an important part in breaking the revolution in St. Petersburg in 1905 and who had been persuaded to stay in service only because of the appeals of Lopukhin. After a variety of disagreements Lopukhin fell from power as director of police of the Ministry of Interior...

The vacillating and weak leadership of the tsar made the other hazards of service hard to bear. All officials faced the possibility of death by assassination. Many began to wonder if the cause—and the leader—were worthy of such sacrifice. Stolypin said he considered each day of his life his last one, and he was unable to think of himself dying a natural death. In 1911 he was shot and killed in Kiev. V.K. Plehvve (whose predecessor as Minister of Interior, Sipiagin, was murdered) told a diplomat that he knew revolutionaries were stalking him, but he could not change his movements lest he show fear in public. A few days later his carriage was blown bup by a bomb and he was killed...

...The regime had gradually become so thoroughly discredited with the public that talented and honest men hesitated to serve it... After a short time in the service of the government, bureaucrats [also]developed an outlook that was anti-noble. Only the Baltic Germans were able to think of themselves as administrators *and* nobles. Because of their peculiar position in which they were permitted to keep their own language, organizations and schools, the Baltic Germans were loyal to the crown rather than to the state, and their position resembled that of the corporations in the old dynastic state rather than that of a racial minority in a modern national sovereignty. They were generally better educated than Russian nobles, and as a result the regime actively recruited them. But this put in the forefront of government ranks a group whose name spelled to most Russians, peasant and noble alike, a heartless efficiency, an unyielding stiffness in administering affairs of state.

...All through history the nobles had been harshly handled by tsars, entire generations of them being wiped out in deadly battles. The last of the great medieval magnates, the Boyars, had disappeared in the early eighteenth century. From the French Revoluition of 1789 until 1861 the tsars and nobles professed love for each other, but in the latter year the autocrat Alexander II sacrificed the interests of nobles on behalf of the state when he freed the serfs and gave them land the nobles considered their own. Since that time the nobility was swiftly descending into oblivion, and the tsars were not interested in doing much to save them. The nobles resented the fact that they had gained nothing from the 1905 revolution, [despite their support of the throne]...Some took the path of the young noble, Vladimir Lenin, who dedicated his life to destroying the regime. Others like Dr. Boris Nikolsky, prominent in gentry organizations and one of the leaders and founders of the Black Hundreds, saw that Nicholas, not the system of government or society, was the enemy and both he and his wife had to be murdered if Russia were to be saved...

Nicholas never clearly understood how he was related to these disaffected groups, nor did he have any idea of how to heal the breach between the autocracy and potential supporters... But he had momentary flashes of insight. In the midst of the first revolution he said to a Minister of Agriculture, "I understand that the position of the government is impossible if it relies only on troops." Five years later he wrote to Stolypin, "The two cornerstones of policy are land for the peasant and migration to Siberia." These remarks brought Nicholas as close to wisdom and as close to the saving of Russia as he was ever able to come. But within two years he was doing much to wreck the political program Stolypin had designed to solve the country's problems. The process in motion... was basic in the political metabolism of Russia... and anything that stood in its way...was bound to be destroyed.

...He had no faith in constitutions or laws as answers to Russia's problems. Early in his reign he thought he found a useful ideology in the teachings of the Easterners, a group that preached Russia was *Kulturträger* and had a great role to perform in the Far East... They believed it was [the tsar's] duty to add lustre to the crown and to enrich and enlarge the state. They said that Manchuria and certain western Chinese provinces along with all Sinkiang, Mongolia, and Tibet could be brought under the sway of the White Tsar, a phrase referring not to a racial policy but to a Greater Russian Empire. "There are no frontiers for us in Asia," said Prince E.E. Ukhtomsky who was a tutor and friend of Nicholas. The annexation of these huge tracts would be an unchallengeable reply of the autocracy to the unresolved political problems. 1904-1905 brought an end to the Easterners. Events of these years [overturned what they] had predicted for the Far East...

At this time Nicholas thought he had a better idea for saving Russia. He proposed the canonization of Serafim of Sarov, a monk who had died seventy years earlier. Serafim had pious qualities that won the admiration of the peasant masses... the minister Plehve espoused the cause, although he did not normally concern himself with religious matters but probably now saw the political importance of the mass following of the *starets*...

Alix was attracted to the cause because he was a *starets*, and he preached one of her favorite articles of faith: those who believed and prayed would always have their prayers answered. In addition it was said that devotion to Serafim would make it possible for her to achieve the thing she wanted most: an heir to the throne. Nicholas did not always share the religious passions of his wife, but he was won over to the canonization because of its political implications. Philippe and Dr. Encausse were strong supporters of the canonization of Joan of Arc, whose case was being actively discussed in Rome and Paris. Those who supported Joan claimed her elevation would exalt the national spirit of France, it would be the weapon traditionalists could use to turn back the assaults of secularism and revolution which at that moment were threatening to win the contest for the soul of France... Nicholas thought he had found another substitute for politics...In January 1903, when he signed the manifesto proclaiming the saint, he remarked that it was a moment of the profoundest satisfaction for him and that it was a memorable day for Russia....an English journalist in 1913 reported that Serafim was already the most popular saint in Russia.

[The church and bureaucracy was attracted] to use the controversy surrounding the canonization as an opportunity to embarrass the regime. Alexander Herzen, the revolutionist, had found a similar opportunity in 1861 to lampoon the government for the elevation of "an unwashed monk," St. Tikhon of Zadonsk... After that, Alexander II avoided raising up other saints. Alexander III had no canonizations for fear of controversy. In 1896 Nicholas induced the church to canonize a former bishop of Chernigov. The new tsar escaped trouble at that time, but in 1903 the public was in a different mood.

...The government was ridiculed because it was accused of devoting all its energies to win recognition of a monk who had sat on a rock in the forest for many years...The church complained that the tsar...was trampling on tradition and the rights of the church... The entire issue of how a saint was proclaimed now came to the fore. Unfortunately truly critical and scientific studies of hagiographical literature had begun only in the 1890's. The church therefore asked the officials to slow down the pace of Serafim's cause, but the wrath of the debate forced matters ahead... In 1901 when the final stage was near, it was pointed out that not enough time had been given to studying the matter. Nicholas proclaimed with some ascerbity that he would give the Holy Synod a year in which to indulge the necessary study and prayerful contemplation. Pobedonostsev realized that the issue was damaging the autocracy, and he also saw the danger in the ecclesiastical church authorities using the issue to talk of an independent church... there might be a call for other free institutions. [This caused him to lose what remaining influence he had]... similar recalcitrance on the part of the influential Bishop Dmitry of Tambov resulted in a sentence to exile... The press of the empire remained indifferent to all these events.

The controversy left behind a heritage of bad feeling. The church suspected that there was something untrustworthy about the faith of the imperial pair. Nicholas was aware that the wounds caused by the affair would not soon heal. The church, the upper classes, and parts of the government had defied him or expressed indifference to a cause close to his heart.

In her letters Alix revealed the extent of her devotion to Serafim, and in her effects found in Tobolsk after her death there were several of his ikons. To her friend Anna Vyrubova she quoted some of the most cherished sayings of Serafim: "The Lord Himself says: All things are possible to him that believeth." Rasputin's presence was justified by, "The Lord hears equally the monk and the simple Christian layman." In one of her last letters from Siberia she quoted Serafim: "When you are reproached—bless; when slandered—rejoice." She reminded her friend that "This is your road and mine." When as a result of her devotion, she believed, a son was born in the terrible summer of 1904 she saw this as a sign and named him Alexis, which means Help of God.

§

During the World War Nicholas tried another project that was supposed to substitute for politics. He ordered the prohibition of alcoholic drinks. The revenue of the state relied heavily on liquour taxes, and peasant drunkenness—always a grave problem in Russia—could be attributed directly to the laws of the state. Nicholas hoped to restore the peasant to sobriety, to provide him with surplus capital that had formerly been squandered on drink, and generally to elevate the moral atmosphere of the country. Needless to say this experiment was not as successful as he had hoped.

§

When Nicholas sought to organize the state around ideas of his own choosing he was clumsy and ineffective, and when he tried his hand at politics he was equally inept. But he stubbornly held on to power... there was at least one political ideal that gave him a feeling of strength. He was king. Although there were only two republics in Europe, many people had the idea that kings were anachronism. Kingship however, had been one of man's most ancient institutions, and its defenders refused to admit that new societies of modern Europe had made kings into epigones. Kingship had adopted itself to all kinds of cultures and types of social organization and had demonstrated viability for millenia. But in 1900 it was hard to believe that the handful of pundits and fanatics who had operated for only a short time could pronounce the death sentence on such a venerable institution. The monarchist who would succumb to such an argument seemed to have no knowledge of history. ∎

From "Lady Macbeth"

...Shortly before the revolution when a large crowd of friendly peasants in the western city of Novgorod greeted her with fervor she said, "What would the babblers of the Duma be able to say about this reception now?" Later, she added, "How they lie, those who assure me that Russia does not love me." ...She felt that somehow the masses were close to her because they shared a common faith with her. She believed, as Nicholas did in the cult of the people. This cult was plain to see at court where many peasant-priests reached high positions. G.I. Shavelsky, who became chief chaplain of the armed forces, was the son of a poor village sexton. Fr. Vasiliev, the chaplain of the court, was a peasant, and so was Theofan, Rector of the St. Petersburg Theological academy, confessor of the empress, and later a bishop.

If she could not approach the people directly she would use an intermediary; this was Gregory's role...[but] his instincts told him to avoid the limelight, to stay in the shadows. His close friend in these early days of his public career, the monk Iliodor...provided him with a ready-made crowd of peasants before whom he could perform. Iliodor had built a fortress-monastery at Tsaritsyn (currently known as Volgograd) on the Volga where he harangued his followers and denounced the abuses of authorities. His actions finally got him into trouble with the Holy Synod and he had been forced to save himself from exile by going to the capital to plead his case with the tsar. Gregory promised to eventually arrange this interview. In responding to the favor, the monk introduced Gregory as a Man of God to the crowds in Tsaritsyn, who were accustomed to believe anything Iliodor told them. In police sent back to Tsarskoe Selo Gregory was described as an idol of the peasants. Later, when Iliodor fell out with his chum he claimed Gregory had tricked him and had been plotting a campaign to get control of the church. Actually, Gregory intended nothing more than to put on a show to convince the empress that he had contact with the peasantry.

...The wife of the government pamphleteer, Gen. Bogdonavich, noted in 1894, a week before the start of the reign, "The new Tsaritsa is not friendly." In the pages of his journal published by the Soviet government we can see the gradual course of deterioration of the relationship of Alix and society. The nasty, wrong-headed gossip increased and the names of more and more members of the court appeared among those who carried the aspersions to Bogdonavich. Even attempts of the empress to cultivate friends misfired and were construed as insults.

The mother of Nicholas II had been able to establish herself as the truly important leader at court during the reign of Alexander III. She was a refreshing difference from the long line of German princesses, and she had genuine abilities as a social leader. Around her gathered the most important people of the social world. She gave them leadership and prestige, and while the emperor preferred to live in semi-retirement, he and his wife appeared often enough in society for the world of the capital to be called brilliant. The court was made up of young and ambitious matrons and their friends. They looked forward to an interesting and glamorous life. At that moment the tsar unexpectedly died. They could not build careers around the next empress; she was supposed to bring in her own friends, "the young court," as distinguished from the "old court."

...Through the first twelve or thirteen years of his reign Nicholas was emotionally dependent on his mother; he looked to her for practical advice as well as consolation... It was hard for the ambitious Alix to accept this situation. Nicholas, Alix, and the Dowager Empress returned from many burial rites of Alexander III and took up residence in St. Petersburg in a common household. Writing to a friend at this time Alix bitterly complained that her husband was blind, he did not see the falseness and hypocrisy of those around him. He spent his days working and in the evenings went alone to visit his mother. She finished, "I am alone most of the time."

...Her first small opportunity to strike against her enemies came as a result of a crisis that occured when the court was in residence in Livadia, in the Crimea, in the summer of 1899. The Dowager, fortunately for Alix, was not present, but many of the ministers were in attendance, waiting for a series of important conferences with the emperor who was ready to discuss with them the situation in the Far East. Present were Sipiagin,

Witte, General Kuropatkin, and the agreeable Count V.N. Lamsdorf who was Foreign Minister at the time. The tsar suddenly became dangerously ill with typhus and the ministers feared he might die. They were at once compelled to face the problem of the succession to the throne. Unless there was a male heir, imperial law was vague on this point. The situation in 1899 was complicated by the knowledge that the empress was known to be in the early stage of pregnancy; the sex of the child became a matter of importance... Alix called in Baron V.B. Fredericks, Minister of the Imperial Court and a member of the Council of State, and told him that in the event her husband should become incapacitated or die, she would become regent. Fredericks expressed the united will of the [cabinet] that the tsar would be replaced temporarily during an illness only by a group of ministers serving collectively as regent. No one believed that a group... could really replace one person acting as autocrat. It seemed to her a trick to make it easier for the government to hand over power to one of the grand dukes. Some favored naming the eldest daughter of Nicholas, Olga, as regent. She would obviously be a tool of the ministers. Witte stepped forward with the nomination of Grand Duke Michael...[who would resign if the unborn child should turn out to be male]. He was to Witte's liking because he would obviously leave the business of government to the ministers while he pursued his pleasures. Witte disliked Nicholas and he later outlined a plot to Lopukhin, the police official, for replacing the tsar with Michael. Alix was alerted to the dangers by the plots which quickly flourished all around her. To her it seemed her husband might be pushed off the throne even if he survived the illness.

To the suprise of all those who were interested in the succession, Alix stubbornly refused to cooperate with the ministers in discussing the matter; she only repeated what she had said to Fredericks. Beyond that she confined herself to long hours of nursing her husband...she was alone, the sole guardian of the legitimate autocracy. Up to this time Nicholas leaned on the support of his mother. However, in the crisis of 1899, the Dowager Empress stood with Witte, her favorite. In court circles it had been rumored for some time that she felt the wrong son was on the throne. She too favored the more personable Michael, especially after Nicholas's disastrous marriage.

The recovery of the tsar in the following autumn put an end to all these discussions. [Alix took the opportunity to remind him she showed more loyalty than anyone else in that dark time, and that most of the ministers and even his own mother] could not be completely trusted. Nicholas was not at once totally won over by his wife's point of view.

Slowly, as the years went on, he began to turn more to her for encouragement. The erosion was under way. When he was well enough to look at some state papers, the ministers noted these were returned to the government departments with marginalia in a strange handwriting—it was that of the empress. From this time she considered herself more than a political helpmate of the tsar; she began to think of herself as the valide, the protectress of the rights of the imperial house. She proposed to conduct the war of defense with the weapons of law and religion. For instance, when Nicholas abdicated in 1917 she was the only one who protested that his act was contrary to a law promulgated by the emperor Paul. In 1909 when she began the construction of the Fedorovsky Cathedral at Tsarskoe Selo, a place where she was to spend many hours every week in prayer, she put in it the household icon of the autocracy. It had been a favorite icon of Ivan the Terrible and in 1613 the nun Martha had used it to bless Michael Romanov on the eve of his coronation.

[Poor health in the first few years of the 20th century again raised doubts about an heir...] She began to think in frenzy of the problem of succession. In this setting Philippe came to court and preached the power of prayer and the cult of Serafim of Sarov whose intercession with God would bring a son to the royal couple. Alix believed him.

...She saw another threat in 1912 when the Grand Duke Michael married a commoner, Natalia Wolfert, daugher of a prominent Moscow lawyer and a member of the moderate Octobrist party. There was a rumor circulating at this moment that the tsarevich Alexis was near death. The rumor was true. Michael had been content with a long liaison that included the birth of a child. Now there was a sudden marraige which would most likely cause the Moscow liberal circles and members of the Duma to favor the duke in the event of another revolution...she was already convinced that the Duma was a nest of vipers and its Octobrist party under Guchkov was preparing plots against the throne... She was outraged by the [implications] and also by Natalia Wolfert's reputation as a court baiter. Nicholas's original anger was increased by his brother's clumsy effort to mollify him... Michael was severely punished by being deprived of all his estates and income and put on a small allowance. Nicholas also refused to grant him the right of succession; a ukase stated clearly that in the event of the death of Alexis, Michael was to be only a regent and not an heir. To the public the vengeful hand of the empress was evident in these Draconian measures. Society was stunned by what seemed only a morganatic marriage. ∎

Appendices

From "*Starets* Ascending"

In 1905 the imperial family deserted the Alexandra Palace and the idyllic world of Peterhof with its parks, statues, waterfalls and cascading fountains. They left this Versailles of the North and took up residence in the Alexander Palace at Tsarskoe Selo, which became their refuge, their Petit Trianon. Most people regarded the village as an unhealthy and dull place. In order to be near court, members of society had lavished money and time on the building and furnishing of handsome villas at Peterhof. Their dwellings at Tsarskoe Selo, however, were modest and uncomfortable. When Nicholas and Alix began to live there they succeeded in cutting themselves off almost completely from society... The place became a strange little world built around the court and the 1500 members who indulged in its endless intrigues and who lived for petty favors or stipends.

§

[Gregory's] lack of interest in office was one of the things that convinced the emperor that he was a man who could be trusted and who had come to court for altruistic reasons. Whatever money he was given came from secret funds provided by the Minister of Interior Khvostov, or from the police official, Beletsky. Tsarskoe Selo was in complete ignorance of this. From 1913 to 1916 rent on his flat in St. Petersburg was paid from court funds. During the war years it was said that he was the ikon-keeper at the palace, but there was no such post at court. Gregory once complained that the empress was so tight-fisted that she did not reimburse him for the money he spent making trips between the capital and the palace.

§

This [the violent encounter depicted on page 127 above] was the end of Hermogen. Shortly after this he went into exile. But he remained loyal to the emperor. Fate gave Nicholas a chance to discover how he misunderstood the bishop when in 1918, in Siberia, Hermogen was murdered by the Bolsheviks who suspected him of participating in a plot to help the imperial family escape.

§

Rasputin's pluck is sometimes evident in these early years. He revealed an ability to see the weaknesses of his opponents and to know something of his own strength. He was the first to note that although Iliodor was a fanatic, he could be controlled by having the promise of a promotion dangled before his eyes. Rasputin did not confuse his actual position at court with the position he bragged about when he was in the salons. He used whatever weapons he had, and they changed with time and place.

...The direction and meaning of Rasputin's career became difficult for his contemporaries to analyze and to understand because parochial issues were entangled in the important questions of the day. As Guchkov once pointed out to his friend, the British historian Bernard Pares, there was something ludicrous as well as shameful about the figure of the "snuffy lay-brother" appearing among the big events of his time. Nevertheless, the question of Rasputin had to be faced. The public refused to pay much attention to other possibly more important events, and there were enough persons at court and in the Duma who saw Rasputin's presence as a burning issue in itself. Consequently, at every turn he was pursued by critics, and he became involved in controversies that centered on problems important to state and church. The result of these adventures was that he acquired an undeserved reputation for vast power. ■

Appendices

From "Politics, War and Revolution"

In the past the autocracy had shown weaknesses in [dealing with matters like food shortages]. In the Crimean War of 1854-55, the Turkish War of 1877-78, and in the famine and plagues of the early 1890's... the public conscience [was beginning to play] a role in national crises. In the disasters of the 1890's newspapers reported and for the first time gave life to the story of a national calamity, incidentally highlighting the government's inability to deal with it. Zemstvos were ready to deal with the problem, but Alexander III believed they were small insidious growths that in time of crisis might turn to democratic organs inside the autocratic state, and in this way become rallying points for revolution. His attitude remained ambiguous. While he feared them he needed them; they were the only institutions capable of organized effort on behalf of the disaster victims. They had organization, dedicated men, and a powerful institutional will.

§

...A memoirist [once] recounted some of Rasputin's remarks about his [reputation]. One evening he walked into a gathering and made a few greetings, then [noticed] someone who might be hostile. At once his mood changed, he became pensive and suspicious. From a bowl on the table he picked up an apple and with one powerful wrench of his hands tore it into halves. He took a large bite and then sat down near his foe. Rasputin's clothes were creased and soiled from a night of carousing. His long hair was uncombed and matted in places. He looked down at the rug covered floor, silently staring for a long time, and then began to speak through a mouth filled with pieces of apple. "Why do you talk bad about me? They say bad things about me in the Duma. It is necessary to speak the truth. One ought not to lie. It is bad, very bad." After a long pause he added "One must always speak the truth." But then he drifted away as if tired of the futile task of trying to defend himself.

§

[Alix liked to invoke Rasputin to remind the tsar of her theory of what the arrival of the *starets* meant, as part of her case that she should have more influence herself]. She frequently reminded Nicholas:

> The wickedness of the world ever increases. During the evening Bible I thought so much of our Friend, how the bookworms and pharisees persecuted Christ, pretending to be such perfection... Yes, indeed, a prophet is never acknowledged in his own country. And how much we have to be grateful for how many of his prayers were heard.

Later she added a plea for Nicholas to "listen to him who only wants your good and whom God has given more insight, wisdom, and enlightenment than all the military put together." ∎

Appendices

From "Ministers Made, Ministers Unmade?"

On 23 November the emperor suddenly informed his wife that he was letting Protopopov go. To soothe her, Nicholas said he was sure that the minister was a good man but he was useless. Then he told her that Protopopov had recently suffered from signs of psychosis. In anticipating her responses to his opinions he warned, "Only, I beg you, do not drag our Friend into this. The responsibility [of finding a new minister] is with me, and therefore I wish to be free in my choice."

...Manasevich-Manuilov was at one time active in manufacturing for the foreign service of the Okhrana certain false plots against the life of the tsar. He also seems to have had some role in one of the most famous forgeries of all: *The Protocols of the Elders of Zion*.

...[because of what he heard from a drunken Rasputin about the Protopopov situation, and what the Grand Duchess heard from Alexandra,] word that Rasputin was the true ruler of Russia spread quickly and it was presumed that the boast referred not only to the present moment but also described a situation that had existed for many years.

...The ministerial changes discussed at length above are, of course, only a few of those in which Rasputin's name was mentioned. The reasons for the selection of these cases has been previously mentioned. In many other hirings and firings he was said to have played a role. The pattern of events is always more or less the same, and while Rasputin's name appears, the appointments and dismissals can usually be explained by causes that have nothing to do with Rasputin. ∎

Bibliography

Abrikossow, D.I. *Revelations of a Russian Diplomat.* edited by G.A. Lensen. Seattle: The University of Washington Press, 1964.
Adam, Juliette. *Impressions françaises en Russie.* Paris, 1912.
Alexander Mikhailovich, Grand Duke. *Once a Grand Duke.* New York: Farrar & Rinehart, Inc., 1932.
Alexander Fedorovna, Empress. *Pisma imperatritsy Alexandry Fedorovny k imperatoru Nikolaiu II.* 2 vols. Berlin, 1922.
Almazov, Boris. *Rasputin i Rossiia, istoricheskaia spravka.* Prague, 1922.
Andreivich, V.K., *Istoriia Sibiri* 2 vols. St. Petersburg, 1887-89.
Anonymous, *Otkrovennye rasskazy strannika dukhovnomu svoemy ottsu.* Paris, 1948.
Anonymous, *Russian Court Memoirs, 1914-1916.* New York: E.P. Dutton & Co., 1919.
Astrov, N.I. and P. Gronsky. *The War and the Russian Government.* New Haven: Yale University Press, 1929.
The Life of the Archpriest Avvakum by Himself, trans. by J. Harrison and H. Mirrlees. Hamden, Conn.: The Shoe String Press, Inc., 1963.
Aziatskaia Rossiia. 3 vols. St. Petersburg, 1914.
Bakhrushin, S.V. *Ocherki po istorii kolonizatsii Sibiri v XVI i XVII vv.* Moscow, 1928.
Beletsky, S.P. *Grigorii Rasputin.* Petrograd, 1923.
Bienstock, J.W. *Raspoutine, la fin d'un régime.* Paris, 1917.
Bogdanovich, A.V. *Tri poslednikh samoderzhitsa. Dnevik.* Moscow, 1924.
Bompard, Maurice. *Mon ambassade en Russie, 1903-1908.* Paris, 1937.
Bookwalter, John. *Siberia and Central Asia.* New York: J.J. Little & Co., 1899.
Boulangier, Edgar *Notes de voyage en Sibérie. Le chemin de fer Transsibérien et la Chine.* Paris, 1891.
Brussilov, A.A. *A Soldier's Notebook, 1914-1918.* London, 1930.
Buchanan, George. *My Mission to Russia.* 2 vols. Boston: Little, Brown & Co., 1923.
Buchanan, M. *The Dissolution of an Empire.* London: J. Murray, 1932.
Buxhoeveden, Sophie. *The Life and Tragedy of Alexandra Fedorovna, Empress*

of Russia. New York: Longmans, Green & Co., 1928.
Cantacuzene, Julia. *Revolutionary Days*. Boston: Small, Maynard & Co., 1919.
Chappe d'Auteroche, J. *Voyage en Sibérie*. Paris, 1768.
Chernov, Viktor. *Rozhdenie revoliutsionnoi Rossii*. Paris, 1934.
Chernovsky, A.A. (ed.) *Soiuz Russkogo Naroda*. Moscow, 1929.
Clark, James M. *The Great German Mystics: Eckhart, Tauler, Suso*. Oxford: Basil Blackwell, 1949.
Combarieu, Abel. *Sept ans a l'Elysée avec le président Emile Loubet*. Paris, 1932.
Curtiss, John S. *Church and State in Russia, The Last Years of The Empire. 1900-1917*. New York: Columbia University Press, 1940.
Dehn, Lili. *The Real Tsaritsa*. London: Thornton, Butterworth Ltd., 1922.
Delacroix, H. *Essai sur le mysticisme spéculatif en Allemagne au quatorzième siècle. Paris, 1900*.
Dillon, E.J. *The Eclipse of Russia*. New York: George H. Doran Co., 1918.
Dmitriev-Mamonov, A.I. and A.F. Zdziarski (eds.). *Guide to the Great Siberian Railway*. St Petersburg: Ministry of Ways of Communication, 1900.
Dostoevsky, Fyodor. *The Brothers Karamazov*. New York: The Modern Library, n.d.
Dzhanumova, Elena. *Moi vstrechy s Gr. Rasputinym*. Petrograd, 1923.
Encausse, Philippe. *Sciences occultes, Papus, sa vie, son oeuvre*. Paris, 1949.
Encausse, Gérard. "Du traitment externe et psychique des maladies nerveuses" (1891): "De l'état des societes a l'époque de la révolution" (1894): "Catholicisme, satanisme et occultisme" (1897): "L'âme humaine avant la maissance et après la mort" (1898): "Comment est constitué l'être humain" (1900). All these pamphlets are in the Bibliotheque Nationale, Paris.
Entsiklopedicheskii slovar (Brockhaus-Efron). 43 vols. St. Petersburg, 1890-1904.
Erman, Adolph. *Travels in Siberia*. Philadephia: Lea and Blanchard, 1850.
Fischer, Raymond. *The Russian Fur Trade*. Berkeley: University of California Press, 1943.
Fülöp-Miller, Rene. *Rasputin, the Holy Devil*. Garden City, Garden City Publishing Co., 1928.
Georgievskii, G.P. *Koronovanie russkikh gosudarei. Istoricheskie ocherk*. Moscow, 1896.
Gilliard, Pierre. *Thirteen Years at the Russian Court*. New York: George Doran, 1921.
Gobron, Gabriel. *Raspoutine et l'orgie russe*. Paris, 1930.
Golovachev, P. *Sibir*. Moscow, 1902.
Gromyko, Maria. *Zapadnaia Sibir v XVII v. Russkoe nasalenie i semledelcheskoe osvoenie*. Novosibirsk, 1965.
Gudzy, N.K. *History of Early Russian Literature*. New York: The Macmillan Co., 1949.
Guérassimov (Gerasimov), Alexandre. *Tsarisme et terrorisme*. Paris, 1934.
Gurko, V.I. *Features and Figures of the Past*. Stanford: Stanford University Press, 1939.

Hiersemann, Karl W. *Originalakten zum Mord Rasputin*. Liepzig, 1928.
de Hornstein, Xavier. *Les grand mystiques allemands du XIVe siècle*. Lucerne, 1922.
Iadrintsev, N.M. *Sibirien*. Jena, 1896.
Jacoby, Jean. *Raspoutine*. Paris, 1935.
de Journal, M.J. Rouët. *Monachisme et monastères russes*. Paris, 1952.
Jundt, Auguste. *Les amis de Dieu au quatorzième siècle*. Paris, 1879.
Kabo, R.M. *Goroda zapadnoi Sibiri* Moscow, 1949.
Katkov, George. *Russia 1917. The February Revolution*. New York: Harper & Row 1917.
Kaufman, A.A. *Krestianskaia obshchina v Sibiri. Po mestnym izledovaniiam 1886-1892 gg*. St. Petersburg, 1897.
Kennan, George. *Siberia and the Exile System*. 2 vols. New York: Century Co., 1935.
Kerensky, Alexander. *The Road to Tragedy*. London: Hutchinson & Co., 1935.
_____. *The Crucifixion of Liberty*. New York: John Day Co., 1934.
Kleinmichel, Countess. *Memories of a Shipwrecked World*. New York: Brentano's, 1923.
Knox, Thomas. *Overland Through Asia*. Hartford: American Publishing Company, 1870.
Knox, Alfred. *With the Russian Army, 1914-1917*. 2 vols. London: Hutchinson & Co., 1921.
Kologrivoff, Ivan. *Essai sur la sainteté en Russie*. Bruges, 1953.
Koon, Kate. *Russian Coronation 1896*. Minneapolis: Privately Printed, 1942.
Korvin-Krasinski, Cyril von. *Die Tibetesche Medizinphilosophie. Der Mensch als Mikrokosmos*. Zurich, 1953.
Koulomzine, A.N. *Le Transiberién*. Paris, 1904.
Kshesinskaia, Matilda. *Dancing in St. Petersburg*. Garden City: Doubleday & Co., 1961.
Kummer, R. *Ein Werkzeug der Juden*. Nürnberg, 1939.
Lansdell, Henry. *Through Siberia*. Boston: Houghton, Mifflin and Co., 1882.
Lantzeff, George. *Siberia in the Seventeenth Century*. Berkeley: University of California Press, 1943.
Laporte, Maurice. *Histoire de l'Okhrana la police secrets des tsars, 1880-1917*. Paris, 1935.
Legras, Jules. *En Sibérie*. Paris, 1899.
Leroy-Beaulieu, Anatole. *The Empire of the Tsars and the Russians*. New York: G.P. Putnam's Sons, 1893-96.
Lescalier, August. *Raspoutine*. Paris, n.d.
Levin, M.G. and L. Potapov, *Narody Sibiri*. Moscow, 1956.
Liepman, Heinz. *Rasputin. A New Judgment*. London: Frederick Muller, 1957.
Lockhart, R.H.B. *Memoirs of a British Agent*. Garden City: Garden City Publishing Co., Inc., 1936.
Maire, Gilbert. *Raspoutine*. Paris, 1934.
Maniguet, Louis. "Contribution à l'etude l'influence des empiriques sur les malades. Etude médico-sociale. Un empirique lyonnais: Philippe," *Fac-*

ulté de Médicine et de Pharmacie de Lyon. Anee scolaire 1912-1920, no. 107.
Marie, Grand Duchess. *Education of a Princess*. New York: Viking Press, 1935.
Marsden, Victor. *Rasputin and Russia*. London: F. Bird, 1920.
Marye, George. *Nearing the End in Imperial Russia*. Philadelphia: Dorrance & Co., 1928.
Massie, R.K. *Nicholas and Alexandra*. New York: Atheneum, 1967.
Meignan, Victor. *De Paris à Pekin*. Paris 1876.
Meyendorff, John. *The Orthodox Church*. New York: Pantheon Books, 1961.
Miliukov, P.N. *Vospominaniia, 1859-1917*. 2 vols. New York, 1955.
_____. et al. *Histoire de Russie*. 3 vols. Paris, 1932-33.
_____. *Rossiia na perelome*. Paris, 1927.
_____. *Istoriia vtoroi russkoi revoliutsii*. Sofia, 1923.
Mosolov, A.A. *At the Court of the Last Tsar*. London: Methuen, 1935.
Müller, Gerhard. *Istoriia Sibiri*. 2 vols. Moscow, 1941.
Murat, Lucien. *Raspoutine et l'aube sanglante*. Paris, 1917.
Nansen, Fridtjof. *Through Siberia*. New York: Frederick A. Stokes, 1914.
Nabokoff, Konstantin. *The Ordeal of a Diplomat*. London: Duckworth & Co., 1921.
Narishkin-Kurakin, Elizabeth. *Under Three Tsars*. New York: E.P. Dutton & Co., 1931.
Nekliudov, Anatoli. *Diplomatic Reminiscences Before and During the World War, 1911-1917*. London: J. Murray, 1920.
Nicholas II. *The Secret Letters of the Last Tsar*. Edited by E.J. Bing. New York: Longman's Green Co., 1938.
_____. *Dnevnik imperatora Nikolaia II*. Berlin, 1923.
_____. *Letters of the Tsar to the Tsaritsa*, 1914-1917. London: J. Lane, 1929.
_____. *Journal intime de Nicolas II*. Paris, 1934.
Oldenburg, S.S. *Tsarstvovanie imperatora Nikolaia II-go*. 2 vols. Belgrade, 1939, Munich, 1949.
Omessa, Charles. *Rasputin and the Russian Court*. London: George Newness, 1918.
Paléologue, Maurice. *An ambassador's Memoirs* 3 vols. New York: George H. Doran Co., 1925.
_____. *Aux portes du jugement dernier. Elizabeth Fédorowna*. Paris, 1940.
Pallas, P.S. *Puteshestvie po raznym provintsiiam rossiiskoi imperii*. Vol. 3. St. Petersburg, 1773-1788.
Pares, Bernard. *The Fall of the Russian Monarchy*. New York: Vintage Books, 1961.
_____. *My Russian Memoirs*. London: Jonathan Cape, 1935.
Petrunkevich, Alexander, et al. *The Russian Revolution*. Cambridge, Mass.: Harvard University Press, 1918.
Poincaré, Raymond. *Au service de la France. Neuf années de souvenirs*. 10 vols. Paris, 1926-1933.

Pokrovsky, M.N. (ed.), *K.P. Pobedonostsev i ego korrespondenty, pisma i zapiski.* Vol. I. Moscow, 1923.
Poliakoff, V. *Tragic Bride.* New York: D. Appleton & Co., 1928.
Polivanov, A. *Memuary.* Moscow, 1924.
Pomus, M.I. *Zapadnaia Sibir.* Moscow, 1956.
Price, M.P. *Siberia.* London: Methuen & Co., 1912.
Purishkevich, V.M. *Dnevnik.* Riga, 1924.
Le Queux, W. *Le ministre du mal.* Paris, 1921.
Radzwill, Catherine. *Rasputin and the Russian Revolution.* New York: John Lane Company, 1918.
Rasputin, G.E. *Moi mysli i razmyshleniia.* Petrograd, 1915.
Rasputin, Marie. *Rasputin.* London: John Long, 1929.
_____. *My Father.* London: Cassell & Co., 1934.
Rodzianko, M.V. *The Reign of Rasputin.* London: A.M. Philpot, 1927.
Rollin, Henry. *L'apocalypse de nôtre temps* (Paris, 1939).
Rosen, R.R. *Forty Years of Russian Diplomacy.* 2 vols. London: George Allen & Unwin, 1922.
Sava, George. *Rasputin Speaks.* London: Faber & Faber, 1941.
Sazonov, S. *Fateful Years, 1909-1916.* London: Jonathan Cape, 1928.
de Schelking (Shelking), Eugene. *Recollections of a Russian Diplomat.* New York: The Macmillan Co., 1918.
Seesholtz, Anna. *Friends of God. Practical Mystics of the Fourteenth Century.* New York: Columbia University Press, 1934.
Semennikov, V.P. *Monarkhiia pered krusheniem. Iz bumagi Nikolaia II.* Moscow, 1927.
_____. *Za kulisami tsarizma: arkhiv tibetskogo vracha Badmaeva.* Leningrad, 1925.
_____. (ed.). *Lettres des grands-ducs à Nicolas II.* Paris, 1926.
_____. *Politika Romanovykh nakanune revoliutsii. Ot Antanty k Germanii.* Moscow, 1926.
_____. *Romanovy i germanskie vliiania,* 1914-1917.
Semenov, A. *Otets Ioann Kronshtadskii.* New York, 1955.
Semenov, P. *Geografichesko-statistcheskii slovar rossiiskoi imperii,* 5 vols. St. Petersburg, 1863-1885.
Semenov, P. *Rossiia. Polnoe geograficheskoe oposanie nashego otechestva. Zapadnaia Sibir.* vol 16. St. Petersburg, 1907.
Semenov, Yuri. *Siberia.* Baltimore: Helicon Press, 1963.
Sergeiff, John (John of Kronstadt). *My Life in Christ.* London: Cassell & Co., 1897.
Shavelskii, Gregorii. V
Shcheglovitov, P.D. (ed.), *Padenie tsarskago rezhima.* 7 vols. Moscow, 1924-27.
Shulgin, V.V. *Dni.* Belgrade, 1925.
Sibirskaia letopisi. St. Petersburg, 1907.
Great Britain, Admiralty, Naval Staff, Intelligence Division. *A Handbook of Siberia and Arctic Russia.* Vol. I. London, 1918.
Simanovich, Aaron. *Raspoutine.* Paris, 1918.

Simpson, James Y. *Sidelights on Siberia*. London: W. Blackwood & Sons, 1898.
Sviateishii pravitelstvuiushchii sinod. *Vsepoddanneishii otchet ober-prokurora sviateishago sinod po vedomstvu pravoslavnago ispovedeniia za 1903-04 gody*. St. Petersburg, 1909.
Smith, Clarence J., Jr. *The Russian Struggle for Power, 1914-1917*. New York: Philosophical Library, 1956.
Smolitsch, Igor. *Leben und Lehre der Starzen*. Köln, 1952.
Sokoloff, Nicolas. *Enquete judiciaire sur l'assassinat de la famille imperiale russe*. Paris, 1924.
Spiridon, Archimandrite. *Mes missions en Sibérie*. Paris, 1950.
Spiridovitch, Alexandre (Alexander Spiridovich). *Les dernièrs années de la cour de Tsarskoie-Selo*. 2. vols. Paris, 1928.
_____. *Histoire du terrorisme russe, 1886-1916*. Paris, 1930.
_____. *Raspoutine, 1863-1916*. Paris, 1935.
Struve, P.B. *Food Supply in Russia During the World War*. New Haven: Yale University Press, 1930.
Taft, Marcus *Strange Siberia*. New York: Eaton & Mains, 1911.
de Taube, M. *La politique russe d'avant-guerre et la fin de l'empire des tsars*. Paris, 1928.
von Taube, Otto. *Rasputin*. Munich, 1925.
Tolmachoff, I. *Siberian Passage*, New Brunswick: Rutgers University Press, 1949.
Treadgold, Donald. *The Great Siberian Migration*. Princeton: Princeton University Press, 1957.
Trufanoff, Sergei. *The Mad Monk of Russia, Iliodor*. New York: The Century Co., 1918.
_____. *The Life of Rasputin*. New York: The Metropolitan Magazine Co., 1916.
Underhill, Evelyn. *Mysticism*. London: Methuen & Co., 1911.
Vasilevskii, I. *Graf Vitte i ego memuary*. Berlin, 1922.
Verstraete, Maurice. *Mes cahiers russes*. Paris, 1920.
Vogel-Jorgensen, T. *Rasputin: Prophet, Libertine, Plotter*. London: T. Fisher Unwin, 1917.
Volkov, Alexis. *Souvenirs d'Alexis Volkov*. Paris 1928.
Vyrubova, Anna. *Memories of the Russian Court*. New York: The Macmillan Co., 1923.
Weber-Bauler, Leon. *Philippe, guérisseur de Lyon à la cour de Nicholas II*. Boudry-Neuchatel, Switzerland, 1944.
Wenyon, Charles. *Across Siberia on the Great Post-Road*. London: C.H. Kelly, 1896.
Whyte, Alexander. *Father John of the Greek Church*. London: Oliphant & Co., 1898.
Wilson, Colin. *Rasputin and the Fall of the Romanovs*. London: Farrar Straus & Co., 1964.
Witte, S. Iu. *Vospominaiia* 3 vols. Moscow, 1960.
Youssoupoff, Felix. *Lost Splendor*. New York: G.P. Putnam & Sons, 1953.

cited articles

Zavzarine Paul P. *Souvenirs d'un chef de l'okhrana, 1900-1917.* Paris, 1930.
Zenkovsky, Sergei. *Medieval Russia's Epics.* New York: E.P. Dutton Co., 1966.
Zhevakhov, N.D. *Vospominaiia tovarishcha over-prokura sviateishago sinod.* 2 vols. Munich, 1923.

Cited by Editor:

Clay, J. Eugene. "Literary Images of the Russian 'Flagellants,' 1861-1905." *Russian History*, Vol. 24, No. 4. Winter, 1997.
Fuhrmann, Joseph T. *Rasputin: A Life.* New York: Praeger, 1990.
Fuhrmann, Joseph T. *Rasputin: The Untold Story.* Hoboken: John Wiley & Sons, 2012.
Highamm Robin D.S, & Showalter Dennis E. *Researching World War I: A Handbook.* Westport, Conn.: Greenwood Press, 2003.
Reilly, Edward. "The First Extant Version of Night on Bare Mountain," in *Musorgsky; In Memoriam 1881-1981*, ed. Malcolm H. Brown. Ann Arbor: UMI Research Press, 1982.

Serials and Articles

Gosudarstvennaia Duma. Stenograficheskie otchety. St. Petersburg, 1906-1917.

Pravitelstvennyi vestnik. St. Petersburg, 1869-1917.

From *Krasnyi arkhiv*: "Dnevnik A.A. Polovtseva," III (1923), 168-89. "Rasputin v osveshchenii 'Okranki,'" V (1924) 270-88. "Pokazaniia A.D. Protopopova," IX (1925), 133-55. "K istorii poslednikh dnei tsarskogo rezhima," XIV (1926), 227-49. "Iz arkhiva Shcheglovitova," XV (1926), 104-7. "Politicheskoe polozhenie Rossiia nakanune fevralskoi revoliutsii v zhandarmskom osveshchenii," XVII (1926), 3-35. "Iz dnevnik A.V.

cited articles

Romanova za 1916-1917 gg," XXVI (1928), 185-210. "Perepiska N.A. Romanova i P.A. Stolypina," XXX (1928), 80-88. "Pisma D.P. Romanova k ottsu," XXX (1928), 200-210. "V tserkovnykh krugakh pered revoliutsii," XXI (1928), 204-213. "Boris Nikolskii i Gregor Rasputin," LXVII (1935), 157-61.

From Byloe: "Telegrammy Grigoriia Rasputina," (no. 5-6, 1917), 228-230. "Nikolaia II i samoderzhavie v 1903," (no. 2, 1918), 190-222. "Pokhozhdeniia Iliodora." (1924), 191-95.

Gourko (Gurko), Wladimir. "L'empereur Nicolas II et l'imperatrice Alexandra Fédorowna I," *La revue universelle*, XLIX no. 5 (June 1, 1932), 566-87.

Hippius, Zinaida. "La maisonette d'Ania" *Mercure de France* (August 1, 1923), 611-662.

Maklakoff, B., "Pourichkevich et l'évolution des partis en Russie," Revue de Paris, V (October, 1923), 721-46.

Romanov, A.F. "Imperator Nikolai II i ego pravitelstvo (po dannym chrezvychaynoi sledstvennoi komissii)," *Russkaia letopis* (no. 2, 1922), 39-58.

Rudnev, N.M. "Pravda o tsarskoi seme i temnykh silakh," *Russkaia letopis*, II (1922), 39-58.

Shumigorskii, A. "Rasputin v sudbakh pravoslavnoi tserkvi," *Istoricheskii vestnik*, (March-June, 1917), 629-38.

Schewaebel, Joseph. "Un précurseur de Raspoutine, le mage Philippe," *Mercure de France,* CXVII (May-June, 1918), 637-47.

Vinogradoff, Paul. "Rasputin," *Encyclopedia Britannica, XXXII* (1922), 249.

"Rasputin," *Bolshaia sovetskaia entsiklopediia*, XXXVI (1955), 61.

"Poslednii vremenshchik poslednego tsara," *Voprosy istorii*, (no. 10, 1964), 117-135.

Martin J. Kilcoyne was born in Harlem, New York in 1922. As a radio operator in the 306th bomber group he sustained an inoperable eye injury from shrapnel that ultimately saved his life: the remainder of his crew were shot down three weeks later. Kilcoyne completed his tour after healing, and returned home to study history and literature. He earned his M.A. from New York University in 1951, and moved to Seattle the next summer to enroll in a doctoral program at the University of Washington. Before and after graduating in 1960, he taught courses on the history of Russia and Chinese communism at the University of Montana and Montana State University, Texas A&M University in Commerce, and the University of Wisconsin in Milwaukee. He settled down south with a family of nine after accepting his final teaching position at East Carolina University in 1969. The position was eliminated four years later without explanation. Dr. Kilcoyne remained in Greenville, NC until his death in 1989.

CPSIA information can be obtained
at www.ICGtesting.com
Printed in the USA
BVHW032321190321
603092BV00001B/190